Praise for *Messy Jobs*

In *Messy Jobs,* Garicano, Li, and Wu bring the discipline of organizational economics to a question too often left to speculation: How will AI actually reshape work? They move past the usual debates about what AI can or cannot do and ask the harder questions. What shapes the incentives to adopt it? How does adoption reshape the incentives to learn? What new configuration of skills will emerge as AI advances? A rigorous, original, and engaging account of how AI will reshape organizations and labor markets, and what it will take to thrive in them.

— **Raffaella Sadun**, *Charles Edward Wilson Professor of Business Administration, Harvard Business School*

Messy Jobs is a brilliant application of price theory. AI changes what is scarce in the economy, and therefore what is valuable. When intelligence becomes cheap, judgment, coordination, trust, and responsibility become more valuable. The authors use this simple, powerful logic to illuminate how AI will reshape work and organizations.

— **Bengt Holmström**, *Paul A. Samuelson Professor of Economics at MIT and recipient of the Nobel Memorial Prize in Economic Sciences in 2016*

This is simply a must-read book if you are interested in the future of work in the age of AI. For decades, Luis Garicano has been a leading voice in how organizations morph and change with new technology and innovation. Together with Jin Li and Yanhui Wu, they have written the definitive text on how AI will affect the labor market. The book is an impressive feat of combining academic rigor with clear explanations and concrete examples. I would recommend this book to anyone interested in learning about what comes next.

— **Alex Imas,** *director of AGI Economics at Google DeepMind, Roger L. and Rachel M. Goetz Professor of Behavioral Science, Economics, and Applied AI, and Vasilou Faculty Scholar at the University of Chicago Booth School of Business*

This is the first book in the AI era that recognizes that most of what organizations struggle with does not involve computational problems. People in messy jobs must hold coalitions together, adjudicate between competing interests, and make change stick. These are political, diplomatic, and interpersonal challenges. As a result, these types of messy jobs will persist well into our AI future. Garicano, Li, and Wu, are neither techno-utopian nor techno-dystopian. They take seriously what machines can do, what humans will do, and how jobs will be rebundled. The economics analysis is lucid and penetrating, and the book pinpoints where human agency will remain paramount. The book is hopeful and practical for anyone charting a career in the coming decade.

— **David Autor**, *Daniel (1972) and Gail Rubinfeld Professor, Google Technology and Society Visiting Fellow, Margaret MacVicar Faculty Fellow, MIT Department of Economics*

AI is not going to lead to mass unemployment, and this is the best book to explain why not. It also illuminates how labor markets are likely to evolve. It is short, to the point, eminently readable, and of extreme relevance.

— **Tyler Cowen**, *Holbert L. Harris Professor of Economics at George Mason University and chairman of the Mercatus Center*

There is a lot of woolly thinking on the topic of AI and jobs. This excellent book contains by far the most thoughtful and economically literate account that has yet been written.

— **Patrick Collison**, *CEO, Stripe*

This book isn't just some economist's armchair theorizing; it's a practical guide. I hope you get as much out of it as I did.

— **Evan Guo**, *CEO of Zhaopin Group, the largest career development platform in China*

Messy Jobs

The Work That AI Cannot Reach

Luis Garicano
Jin Li
and Yanhui Wu

Messy Jobs: The Work That AI Cannot Reach

Published by Upriver Press
P.O. Box 51455
Colorado Springs, CO 80949
upriverpress.com

ISBN Print Version: 9798995221906
ISBN Ebook: 9798995221913

Cover Design: James Clarke (jclarke.net)

Printed in the United States of America

Messy Jobs

The Work That AI Cannot Reach

Luis Garicano
Jin Li
and Yanhui Wu

upriver
PRESS

To our kids and all those who will start their working lives in this new world. Our concern for their future motivated us to write the book.

Contents

Preface

In February 2026, the head of Microsoft's AI division told the *Financial Times* that most white-collar tasks could be "fully automated by an AI within the next twelve to eighteen months." He described the jobs in question as "sitting down at a computer, either being a lawyer or an accountant or a project manager or a marketing person."[1] The chief executives of Anthropic and OpenAI have made similarly apocalyptic claims.[2]

These leaders base their views on the rapid shift in their large language models (LLMs), which have evolved from generating text to carrying out tasks with little supervision. This new technology, which for simplicity we call *artificial intelligence*, suggests a future when you might wake up to a message like this from your AI agent: "Good news. I changed your internet provider and cut your internet bill by 10 percent. I also found, booked, and paid for a summer house that fits your needs."[3]

There is little question that AI agents will reorganize large parts of economic life. But predictions about the impact of AI are too simplistic. Think about life after you read a message from your agent like the one above. Your spouse may have changed the plans the previous night, or your flatmates didn't do the dishes as agreed, or your children may have decided that they are *not* going to their swimming lessons.

We are all managers of our own tiny, fast-moving organizations. Your household is an economy. You need to decide how to allocate scarce resources (time, attention, and money), figure out the division of labor (who cooks and who does the dishes), and negotiate agreements such as, "I'll pick up the kids on Tuesdays" and implicit ones like, "I will not read my WhatsApp messages during dinner." The bottleneck is not limited to making sure all decisions fit with each other when circumstances change; it also involves the "politics" of the household, such as making sure the decisions are acceptable to everyone and then implemented. Your family life, like everyone's, can be a mess.

This messiness is why, in our view, doomsday predictions about white-collar jobs disappearing are wrong. AI will certainly have an impact, but the people making the predictions do not understand what most white-collar workers do all day.

All knowledge work varies along one dimension: messiness. At one end of the spectrum, there is a single defined task to execute, such as helping clients file their taxes. You get the pay slips and expenses via email, use rules to fill out a form, and get a result. At the other end of the spectrum, there is an enormous bundle of complex tasks. Running a factory, or a family, involves many different tasks. They are very hard to specify in advance, and they are difficult to split from each other so that some can be automated autonomously. Many involve a lot of conflict.

Along the messiness spectrum, AI has varying degrees of ability to help or replace humans. It is easy for AI to replace people who do simple, clean tasks, but it is hard for the technology to replace workers who do the messy jobs. That does not mean that people who do messy jobs are immune from change; AI could still *reshape* their work. That's because AI unbundles some jobs and rebundles others.

AI technology is changing rapidly. Managers are worried, workers are scared, parents are concerned, and policymakers are confused. This book is designed to help everyone make better decisions and discover new opportunities. We assume that *agentic cognition,* which refers to the ability of AI agents to perform tasks autonomously, will become increasingly cheap and widespread. AI will be able to use all digital tools, but we expect autonomous learning and physical embodiment to remain limited.

The future is not shaped by technology alone. Understanding the consequences of AI requires economic reasoning about scarcity, about complementarities and bottlenecks, about direct effects versus equilibrium responses, about signaling and incentives, and about the organization of work. These are the economic concepts we use to understand the future. Our specific examples will age, but these concepts will not.

London and Hong Kong, May 2026.

CHAPTER 1

Cheap Cognition

The Autonomy Threshold

In January 2026, a student approached one of us and claimed that, with the help of new AI tools, she could complete a PhD in ten days instead of four years. She had a point. Tools such as Claude Code allow a researcher to use AI agents to download data, find relevant literature, conduct statistical analysis, craft a formal model, and even write a full paper. If we only want to produce a standard piece of work, the student would be mostly right.[4]

However, if the student wants to obtain a PhD in order to advance science as much as possible and to later have a good job, then using AI to complete the PhD does not necessarily shorten the timeline. If the tool makes competent performance easier, then more people can do the work. That increases the effective supply of labor. So, the question for our student is not just whether she becomes more productive; the question is, what remains *scarce* after everyone uses the same tool.

This anecdote illustrates the crucial difference between productivity and wages. That difference is driven by scarcity. We see this economic principle at work in history. Previous waves of technological change created different scarcities. Understanding that history can help us understand what is coming. Two precedents teach different lessons: the Industrial Revolution and the information technology revolution.

Two revolutions

The Industrial Revolution: technology eventually helps all

Before the Industrial Revolution, apprentice weavers spent years learning their trade. A master transferred skills by training someone for more than a decade. The best clothes would sell at high prices, because few people could produce high-quality work. Skill and experience were scarce.

Then machines arrived. The Edmund Cartwright power loom, invented in 1784 and widely adopted by the 1830s, meant that a teenager could learn to handle the loom in days and produce in one day a volume of cloth that would have required weeks from a master artisan. The skilled Lancashire weavers faced ruin as their cumulative experience became almost useless.[5]

However, after roughly sixty years of stagnating wages—from about 1790 to 1850—something unexpected happened that pessimistic observers like Friedrich Engels and Karl Marx had not predicted: The sons and grandchildren of the weavers did not remain poor. By the end of the nineteenth century, the real income of most workers had doubled. Machines made workers' lives better.[6]

How did this happen? As the number of machines per worker grew, each worker could produce more cloth. The machine could do the heavy work, but the worker remained necessary. Somebody had to put the cotton in and make sure the machine continued to operate. The worker was complementary to the machine.

Why did workers, rather than capital owners, capture the gains? Institutions, such as unions and democracy, helped to increase workers' bargaining power.[7] But what was critical was the relative scarcity of capital and workers. The supply of workers was limited, but capital, in the form of machines and factories, could be built and reproduced. With enough time and money, anybody could build another factory or another textile mill. As the earnings of capitalists went up, investors built more capital equipment. Competition persisted until the returns to capital went back to normal. Workers were the scarce factor. Hence, workers eventually benefited.

The IT revolution: technology helps highly skilled workers

Compare what happened during the Industrial Revolution with the information technology (IT) revolution. Mass-produced personal computers, like the Apple II in 1977 and the IBM PC in 1981, spread into factories and offices. Like the mechanical loom, they automated work that previously required human skill. But their impact on wages was very different from what occurred previously. In the United States, from 1980 to 2020, the real median weekly earnings of full-time workers with a bachelor's degree or higher increased by about 20 percent, while the earnings of workers with a high school education declined slightly.[8] The earnings gap between college-educated and high-school-educated workers, what economists call the *college wage premium*, nearly doubled.[9]

Why did the IT revolution not follow the same pattern as the Industrial Revolution? Because the IT revolution generated a different kind of scarcity. The mechanical loom complemented all workers. Anybody could learn to use it in a few days. By contrast, computers helped skilled workers more than unskilled workers. To use a database or analytical software, workers needed high-level cognitive skills.

Computers made knowledge *acquisition* cheap, but they placed knowledge *utilization* (the hard work of applying knowledge to solving real problems) entirely in human hands. The result was that computers amplified the economic value of people with higher levels of education. A financial analyst with a spreadsheet was worth far more than an analyst without one, but only if the analyst knew what to model. A lawyer with a database of case law could conduct research faster, but the technology only helped the lawyer who knew which precedents mattered. Technology complemented knowledge; therefore, the workers who gained most from the technology were those who had high levels of knowledge and education.

As a result, information technology divided the labor market. The relative demand for highly educated workers who could use computers as tools went up. The relative demand for office workers, accountants, and administrative assistants, whose routine cognitive work was substituted by computers, dropped. The relative demand for workers at the base of the distribution, such as janitors, health assistants, and restaurant workers, and others doing physical labor without computers, stayed flat. The result was economic polarization.[10]

The two revolutions, industrial and information, led to distinct outcomes. The Industrial Revolution was, eventually, good for all workers because the machines complemented all the workers. The IT revolution was very good for highly educated workers, but far less beneficial for many others.

Which historical precedent is more relevant to the AI revolution that is now taking shape? If AI is like the power loom, then we can expect a painful transition followed by broadly shared prosperity. If AI is like the personal computer, complementing only those who already have scarce skills, then economic polarization will increase.

The AI revolution could create a third possibility. The IT revolution cheapened codifiable information processing, but it required human cognition to operate it. What if AI can perform many codifiable cognitive tasks by itself? That would make some forms of cognition abundant, and it would shift cognitive scarcity toward other human traits.

To understand this new technology, we propose a new framework. Central to this framework is what we call the *autonomy threshold.*

The autonomy threshold

Adaptive cruise control helps you drive. It adjusts your speed to match traffic, maintains safe distances, and handles acceleration and braking. It makes you a better and safer driver. But it doesn't significantly change the economics of driving because the human must remain in the driver's seat. Human time is still the limit on how much driving gets done in the economy. By contrast, a fully autonomous vehicle transforms the economics entirely. The car no longer needs a human, so the supply of driving services is no longer limited by human time. Driving is limited only by machine capacity.

The autonomy threshold is the boundary between machines that *complement* humans and machines that *replace* humans. This is the point at which AI switches from assisting humans to operating alone. Consider what happens if AI can deliver a finished output whose quality, at a given cost, makes hiring a human to check, correct, or own the result too expensive.[11] This boundary is the autonomy threshold.[12]

The autonomy threshold applies to tasks and jobs, not to workers. Workers hold jobs, and jobs involve many tasks. A worker's exposure to AI replacement depends on how many of her job's tasks cross the autonomy threshold. Her future also depends on whether the remaining tasks are discrete and therefore can be cleanly separated from each other (unbundled). By contrast, the job will be safer if it includes tasks that form a "strong" bundle. In this case, peeling away the automatable tasks is difficult, as we discuss in chapter 3.

Below the threshold, AI operates in assistance mode. It helps humans but people remain in control of the finished product. The human reviews the work, catches errors, and takes responsibility for the outcome. In this mode, the economic outcomes are similar to the outcomes of the IT revolution. Employers need fewer hours of human time per unit of output, and that expands the workers' capacity. The analyst who typically writes two reports a week will now be able to write many more in the same amount of time.

When AI crosses the autonomy threshold, when it can deliver a finished solution without a human who checks and owns the outcome, then everything changes. Below the threshold, AI complements humans. Above it, AI can replace humans who do that task. Even when that occurs, AI will not necessarily replace the whole job or the worker.

Three tiers of the autonomy threshold

Every job sits at a different point on the messiness spectrum, depending on the autonomy thresholds of its component tasks and the cost of separating them. A job consisting of a single, well-defined, verifiable task has a low threshold and will fall quickly to AI. A job whose threshold is high, such as one that comprises a tangle of interdependent tasks involving people and politics, is messy. It will resist automation even when the technology improves enormously. The question for any worker is: How high is my job's threshold and how fast is the technology climbing toward it?

Two things make a job messy. First, the tasks that compose it are entangled and hard to separate. This could be because partial outputs are difficult to verify, because reallocating components is costly, or because significant spillovers run across tasks. Second, the job could require holding together a complex nexus of human relationships. This inevitably involves managing conflict or mediating opposing interests. It also involves navigating people's personal lives, moods, and agendas (e.g., our parenting example in the preface). This second source of messiness is far harder to automate, as we shall discuss.

The following figure conceptually depicts the relationship between automation and messiness. On the horizontal axis we plot increasing messiness. On the vertical axis we plot how capable AI is. The curve depicts the increasing autonomy threshold as messiness increases. Very capable AI becomes less likely to replace a job as we move from clean, single-task jobs (tier one) to messy bundles of tasks (tier two) to a nexus of human relationships (tier three).

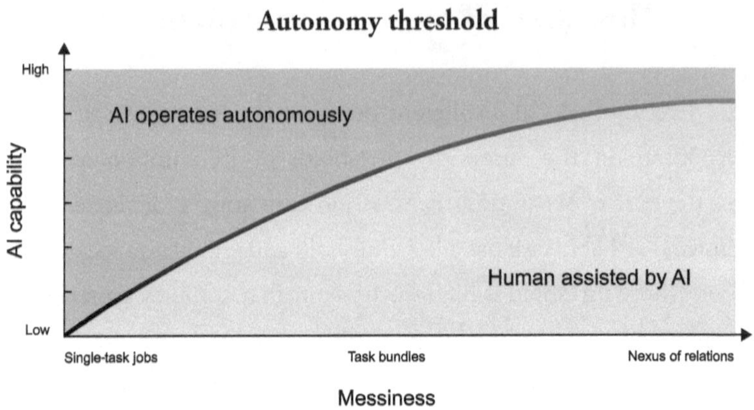

Tier one: single-task jobs

On the left side of the figure are jobs that consist of a single task or a narrow cluster of closely related tasks with verifiable outputs: technical translation, coding a predefined problem, classifying images, and drafting a contractual clause. These are clean problems; they have clear inputs, well-defined success criteria, and verifiable outputs.

For these jobs, the autonomy threshold is low. Currently, AI is already crossing or has crossed the threshold for some of them. Answers can be verified as correct, and hence we can train the system on the right answers, measure its performance, and improve. Once the threshold is crossed, the human worker has little to sell. For instance, researchers found that job posts for writing and coding on a major freelancing platform fell by 21 percent relative to labor-intensive jobs within eight months of ChatGPT's launch.[13] That does not, however, mean human income in the sector necessarily goes to zero, as we will see.

Tier two: bundles of tasks

Tier-two jobs consist of multiple tasks bundled together inside a single role. A medical doctor does not just diagnose illnesses. She also triages and communicates with patients, consults with colleagues, trains residents, and bears legal responsibility for diagnosis. AI may cross the autonomy threshold for some diagnostic tests while remaining below it for other tasks in the bundle.

In tier two, the autonomy threshold for the whole job is much higher than the threshold for any individual task within it. All the tasks in the bundle require knowledge. Some tasks, such as diagnosis, only need codified knowledge, whereas others need contextual, interpersonal, and local knowledge. The former tasks are easy for AI to replace, but the latter ones are difficult to automate.

In many instances, splitting the tasks apart can also fragment the knowledge that is relevant to job success. For example, a radiologist who reads a scan already knows the patient's history, already has a relationship with the referring physician, and already faces the legal obligation that goes with providing a diagnosis. Unbundling the scan-reading task and handing it to an AI program would require someone to reproduce the whole context separately. This would create a coordination cost that might exceed the potential savings from unbundling the task and handing it to AI.

The characteristic structure of tier two is a single human being supported by the machine, either from above or from below.[14] In the "robots below" configuration, AI takes over the routine problems of the job while the human handles the exceptions. An entrepreneur with AI tools can build a prototype in hours that previously required a team of engineers. But it is the entrepreneur who discovers the problem, creates the solution, chooses the market, and bears the risk. AI increases the leverage of a human's exceptional knowledge.

In the "robots above" configuration, AI handles the more sophisticated aspects of cognitive work and deals with the exceptional problems that are unusual or hard. A nurse practitioner with AI diagnostic support still examines patients, communicates diagnoses, and deals with emotional and physical tasks. However, in this case, AI complements the nurse because she does not need to incur the cost of acquiring more medical knowledge to solve a wide range of difficult problems.

Autonomous AI harms workers who perform simple tier-one tasks, but it may help workers who perform a bundle of tasks. AI may help highly trained experts when it removes some of their routine work. And AI may help workers with less training by giving them access to expert cognition without needing to acquire it themselves.

Tier three: a nexus of relationships

In the third tier, depicted on the right side of the figure's horizontal axis, are the jobs where the autonomy threshold is so high that almost all the work requires a human. For example, we know the head of engineering at a manufacturing plant who must decide whom to hire, which machines to buy, and how to install them in the plant. She must negotiate with the workers and the higher-ups on proposed solutions, and she must mobilize the resources to implement those solutions. That bundle of tasks is extraordinarily hard to automate. AI might be able to draft her reports, optimize her schedules, and analyze her data, but none of those tasks are complicated. A lot of her work involves managing friction with other humans and institutions. She must work with the town hall to obtain permits and with people at corporate headquarters to secure funding. Knowing who to ask and how to ask involves tacit knowledge that is not available in AI training data.

In tier-three jobs, managing interdependence between humans, not just between tasks, is messy. There must be a human being who holds the nexus of relations together. The whole job must be performed by a human who handles the interdependencies, manages the friction, and is accountable for the result. As Boris Cherny, creator of Anthropic's Claude Code, put it in February 2026, "Someone has to prompt the Claudes, talk to customers, coordinate with other teams, decide what to build next. Engineering is changing and great engineers are more important than ever."[15]

Tier-three work is different because its messiness is not a temporary limitation on AI's capacity; it is a permanent feature of the problem. Organizations are coalitions of people with legitimate but conflicting goals. To function well, they require someone with the authority to settle disputes and manage those relationships. In addition, someone will need to do even messier work: redesigning entire organizations so AI can be deployed at scale. Neither problem becomes cleaner when computers get faster.

Relationships also matter on the market side. Some jobs survive AI not because the task is technically beyond automation, but because the customer values the human behind the task. Examples include a live concert, a handmade card, or a sommelier in a fine restaurant. In each case, a machine may be able to produce the same objective result, whether it is a sound, a print, or a recommendation. But automation changes the product because it removes the human origin that the buyer was paying for. Chapter 8 examines this demand-side threshold, the point at which buyers no longer treat a machine-made output as equivalent to one with human origin. A job survives when either the supply-side threshold is high enough or the demand-side threshold is high enough.

If relationship-dependent work is so costly, could firms and markets reorganize to need less of it? Chapters 6 and 7 argue that relational work is a structural feature of organizations composed of people with conflicting interests, and of markets where consumers buy not only outputs, but also the human source behind them. Some relationships cannot be reduced to contracts, prices, or algorithms.

This same limit shapes what AI can do once inside the firm. As part III shows, AI does not only change which jobs survive; it changes the systems that make work trustworthy. It cheapens what spot markets already do well: search, contracting, execution, and the verification of standardized output. It does not cheapen work that depends on long-term relationships: reputations, trust among people who expect to keep working together, quality standards backed by someone willing to defend them, and apprenticeships. Chapters 9 through 12 show how these systems must be rebuilt around the relational mechanisms that AI cannot supply. Firms, even when restructured to accommodate AI, remain the most common home for messy work.

The messy demand: satiation versus the Jevons effect

The discussion above was mostly about what AI can do; that is, the supply side. But understanding the market value of workers also requires understanding the demand side. AI makes cognition cheap, and that reduces the cost of goods that require intensive cognitive labor, such as software or legal documents. Cheap cognition makes it possible to solve many previously unsolvable problems. Hence, we expect demand for those cognitive tasks to increase. In broad terms, there are two cases (or scenarios), depending on the size of that increase. We call one the *satiation case* and the other the *Jevons effect.*

Satiation. In 1850, when nighttime fell over London, the city was transformed. The wealthy burned candles that cost more per hour than a laborer's daily wage. The poor huddled around dried plant stems dipped in animal fat, which provided barely enough illumination to prevent stumbling into furniture. Then, over the next century and a half, as Nobel Prize-winner William Nordhaus documented, the cost of light fell by a factor of forty thousand.[16]

Today, the illumination industry has lost most of its economic value. Illumination expenditure fell from a significant part of the budget expenditure to a rounding error. Why? Humans do not demand forty thousand times more light just because it is forty thousand times cheaper. Once we have enough light to see well, we use our money to buy other things. At the risk of exaggerating, the innovation of electric light was so successful that now we barely think about the economic importance of illumination.

The same is true for agriculture. In 1900, American farms

employed 40 percent of the total labor force. Since then, farming has become immensely more productive. [17] It is so productive that the agriculture sector employed only 1.2 percent of the labor force in 2022, which is fewer people than the pet care sector.[18] Agricultural employment collapsed even as output increased because of limited demand. That is another example of satiation.

In economic terms, demand becomes *inelastic* when consumers are close to satiation. This could well be the case with AI. If demand for cognition-intensive goods becomes inelastic, then the tasks that AI is automating will become economically irrelevant. Humans will then have to make their living in other sectors that are less affected by AI. Chapter 8 discusses those sectors, including artisan production, personal services, and musicians who perform live music.

The Jevons effect. Satiation is not the only possibility. If demand dramatically *increases* as the cost of producing cognition drops, then employment in automated sectors will also increase. This is called the Jevons effect, named after nineteenth-century economist William Stanley Jevons. He observed that as steam engines became more efficient, total coal consumption did not decline as some expected; instead, coal use increased significantly.[19]

It is too simplistic to say that AI will make it less costly to find solutions to more problems, or that, as a result, demand for workers will increase. That's because AI itself will deal with some of the problems. Human employment will increase only when humans remain complementary to newly expanded AI ability. In other words, there is a step from the Jevons effect for output to the Jevons effect for employment. If new demand is directed to tasks above the autonomy threshold, AI will fill it. Employment expands only when new demand requires tasks below the threshold; that is, tasks where humans remain the bottleneck.

Consider the impact of electronic spreadsheets in the early 1980s. They automated the main task of financial analysis. Calculations that took days to complete with pencil and paper now took minutes. One would have expected the technology to decimate the jobs of financial analysts. Instead, the demand for financial analysis exploded. Firms could now model scenarios they had never bothered to consider and update forecasts weekly instead of quarterly. The spreadsheet automated the arithmetic but created enormous new demand for doing elaborate calculations and understanding what the numbers meant. The analyst was complementary to the tool. Jevons won.

In both the satiation and the Jevons cases, there will be jobs to do and ongoing demand. In the first case, which we discuss in chapter 8, the automated sectors will lose economic weight and demand will move toward sectors that are least susceptible to automation. In the second case, demand grows so much in automated sectors that employment increases within them.

Road map

Part I focuses on the messiness spectrum. We start with single-task jobs where the threshold is lowest (chapter 2). Then we examine how bundling tasks into a single role raises the threshold (chapter 3). In chapters 4 and 5 we show how AI reshapes the knowledge hierarchy inside those bundles: from below (empowering experts) and from above (empowering non-experts).

Part II addresses the messiest end of the spectrum: the

organizational politics and implementation challenges that no algorithm can handle (chapters 6 and 7), and the demand-side premium that protects jobs even when AI could technically do them (chapter 8).

Part III turns to how organizations must redesign themselves when more work becomes a human-AI bundle. The topics include: screening workers when the old signals collapse (chapter 9); maintaining quality when verification costs explode (chapter 10); setting standards when the production function changes shape (chapter 11); and rebuilding the training pipeline when apprenticeship economics break down (chapter 12). Each chapter also derives the career implications for workers.

The conclusion widens the frame from jobs to the overall AI economy. It shows how the *Messy Jobs* framework helps to explain changes in production, the continuing role of firms, the reshaping of markets, and the limits and possibilities of AI-driven economic growth.

Part I

The Messy Jobs Spectrum

CHAPTER 2

Tier One: Disappearing Single-Task Jobs

When AI Crosses the Threshold

Andrej Karpathy, a founding member of OpenAI and former AI director at Tesla, is one of the most capable programmers in the world. So when he announced in February 2026 that he had stopped writing his own code, people took notice.[20] His remark captured fundamental changes in what programming had become.

When tools like GitHub Copilot first appeared, it was celebrated as a powerful assistant. It could predict the next few lines, fill in boilerplate forms, and save hours of typing. But the basic structure of the work remained. Humans still designed the logic, checked the output, and stayed firmly in control.

But within a few years, the technology changed the workflow. Karpathy coined the term *vibe coding* in 2025 to describe a radically new way of coding. Instead of writing code line by line, a user could simply describe her intent in plain English and let the machine do the job. "I fully give in to the vibes and forget that the code even exists," Karpathy explained.[21]

AI had evolved from helping a human write code to doing the

coding entirely on its own. According to Nvidia CEO Jensen Huang, "It is our job to create computing technology such that nobody has to program. Everybody in the world is now a programmer."[22]

For many programming tasks (not the ones Karpathy does!), AI is crossing the autonomy threshold. These are tier-one jobs: single tasks with clear inputs, verifiable outputs, and low autonomy thresholds. This chapter examines what determines when that threshold is crossed and what happens to the workers on the other side.

Autonomous AI

Coding is often seen as being highly exposed to AI.[23] But other sectors are also susceptible to AI disruption. Consider, for instance, the evolution of the customer support industry over the last three years.

In a recent paper, economists Erik Brynjolfsson, Danielle Li, and Lindsey Raymond analyzed a 2023 experiment in which a Fortune 500 software company gave AI assistants to its 5,179 customer service agents.[24] They found that AI did not help the best agents, but it helped the least experienced people increase their productivity by 35 percent. Novices were suddenly solving problems at rates similar to veterans with more than a year of experience. The AI-based assistant had absorbed the best workers' call transcripts and then gave novices access to expertise that normally took months to acquire. Across the whole call center, the gains weren't limited to the least experienced workers. Average resolutions per hour rose by about 14 percent, agents handled more chats, and customer satisfaction increased slightly. AI was effectively assisting customer support agents.

Contrast that outcome with the experience of the call center at Yum China, one of the largest restaurant companies in China. The company operates more than eighteen thousand KFC, Pizza Hut, and other restaurants. Between 2018 and 2025, as digital ordering grew, customer contact volume increased sixfold to more than sixty million per year. To deal with this, Yum China deployed C-Smart, an AI agent that evolved from simple keyword matching to advanced generative AI.[25]

Without human intervention, C-Smart agents can fully resolve simple customer requests, such as checking order status, asking for faster delivery, or inquiring about restaurant locations. C-Smart uses human agents only for complex or emotionally charged issues, such as complaints and disputes, which account for about 10 percent of the cases. AI has crossed the autonomy threshold for the simple tasks. This allows Yum China to handle six times the volume with roughly a third of the people.

Full autonomy, as we defined in chapter 1, marks the point where a human's contribution to the task no longer justifies the cost of including her. When AI is not fully autonomous, for whatever reason, the process works better with her than without her. She catches errors, supplies context, and provides trust to the buyers. AI is a complement to her efforts. After AI attains full autonomy on a particular task, her contribution becomes less valuable. The AI will replace the human on that task. If the job consists entirely of that task, as is the case in tier one, the human will have little or nothing to sell.

Consider the case of Olivia Lipkin, the sole copywriter for a tech startup in San Francisco. Her role required her to define the company's voice. In late 2022, her assignments dwindled. Then, on the company's internal Slack channels, her managers began referring to her as "Olivia/ChatGPT." It seems that the choice was not between her and another coworker but between a salaried employee and a zero-cost large language model. In April 2023, Lipkin was let go without explanation.[26]

The determinants of the autonomy threshold

The point at which AI can reach full autonomy is not determined by technology alone. In practice, autonomy is often limited less by raw capability than by whether an organization can safely delegate authority to a system that will sometimes be wrong. Three forces determine when that happens for any given type of work: technology, regulation, and consumer preferences.

Technology

The first determinant of the autonomy threshold, technical capability, refers to when an AI system can produce correct solutions to a wide range of real-world problems. A model might perform well in tests, but it can still fail when it is deployed if the real work depends on local context and tacit knowledge that never made it into the training data. If the AI cannot deal with real-world disorders, then the human must remain in the loop to correct its mistakes.

The technology is improving rapidly. Many skills that looked impossible for AI one or two years ago, such as doing high-level math and writing complex prose, are now trivial for today's models. An important empirical predictor of whether today's models can handle a task autonomously is the time it takes for a skilled human to complete the work. Current models handle short, clearly defined tasks well, but they suffer as the time horizon expands and as the tasks require sustained attention, context, and error correction.[27] Researchers have described this as a "jagged technological frontier." The frontier appears within a job when AI dramatically improves performance on some tasks and actively harms it on others.[28]

Three obstacles to technological improvement are fundamental. They impose essential constraints on what AI can do, and there is uncertainty within the scientific community about how fast they will improve.[29]

The first obstacle to improvement is task *verifiability*. For verifiable tasks, AI has a target, like writing a correct piece of code, predicting the shape of a protein, or making a good move in the game of Go. In these cases, AI can learn from its mistakes and improve rapidly. AI is likely to remain imprecise for tasks whose correctness is subjective.[30] This type of work might include writing a novel, designing a logo, choosing a restructuring plan, or deciding how to balance the aggressiveness of a cancer treatment versus the quality of life in a clinical setting. Here, it is harder to apply learning techniques, so AI is likely to remain imprecise.

The second barrier is that most current AI systems do not learn from deployment in the way human professionals learn from experience. A human professional accumulates better judgment over years of working on a variety of cases (patients, clients) during their careers. By contrast, the factors (or criteria) used by an AI model to make "judgments" are usually fixed until it is trained again. Any memory it has is typically an external product feature.[31] Hence, AI cannot build on yesterday's work the way a junior associate can gradually become a senior partner. This matters most in roles where judgment improves over time.

Anthropic's Amodei has offered a partial counterargument to this concern. As these systems learn to process vastly larger volumes of information at once (as context windows expand), a model may be able to hold an entire career's worth of case files, patient histories, or project logs. This could approximate something like human "experience," by giving AI systems "context." But the provision of generous informational context is not the same as learning.[32]

The third obstacle is that AI has no body and no physical presence. This sounds trivial until you consider how much economic life depends on someone being *there*. A nurse holds a patient's hand to reassure her, or a manager walks a factory floor and notices what the reports leave out. Robotics is advancing, but the gap between generating text and navigating an unstructured physical environment is vast. It is unclear how rapidly that gap will close.

There are two other *operational* obstacles. Many people believe they will be overcome relatively quickly, but they still raise the autonomy threshold.

First, AI responses can be biased. Given how LLMs are trained, they are likely to remain so. Large language models are trained, in part, by asking humans which of two responses they prefer. Researchers at Anthropic showed that this training process rewards agreement more than factual accuracy. When an AI response matched the user's opinion, the user was more likely to prefer that response even when the AI response was incorrect. Humans, and the reward models used to train AI assistants, often preferred convincingly written sycophantic responses more than correct ones. The result is that models learn to flatter users.[33]

Other research has found that models become more conformist as a dialogue progresses. The longer the exchange, the more the model drifts toward the user's position, even when that position is wrong.[34] And when a user pushes back with a persuasive but erroneous counterargument, models frequently abandon the correct answer and adopt the user's mistaken view. Researchers call this "stance fickleness."[35] A lawyer who asks a model to find support for a weak legal theory may receive exactly that: confident, well-cited support for a position the model should have questioned.

AI's mistakes are not random. They are systematically biased toward telling people what they want to hear. This matters for AI autonomy because the bias is invisible to the user. A system that tells you what you want to hear will pass your review, precisely because you want that outcome. The human-in-the-loop check that is supposed to catch errors becomes unreliable when the errors are designed, by training, to evade the checks.

Second, AI autonomy often fails at acting safely inside real systems. Once an AI system has real authority and can take actions, small mistakes can become large incidents.[36] Consider Amazon Web Services. In mid-December 2025, AWS engineers allowed Kiro, Amazon's own AI coding tool, to make changes to a production system. The tool, designed to act autonomously on behalf of its users, decided the best course of action was to delete and recreate the environment. The result was a thirteen-hour outage of a customer-facing system. The AI agent was sophisticated enough to write and deploy code, but it was unable to "understand" that deleting a live production environment would be catastrophic. That is the jagged frontier.[37]

Regulation

The second determinant of the autonomy threshold is the law. For example, regulations determine who is permitted to sign and certify contracts, and who is liable for the consequences. This force is often stronger than technology.

At the start of the twentieth century, operating an elevator was a career. You needed some skill to stop the elevator in the right spot. By 1930, automatic elevators with a button for each floor were technically flawless. They were safer and cheaper than a human-operated elevator. But elevator operators still had a job in the US for another twenty years.[38]

Two forces kept them employed: consumer trust and politics. On the one hand, the public initially did not trust automatic elevator brakes. On the other hand, unions fought hard to keep the operators employed. As a result, cities passed laws mandating a human operator to ensure "safety." The technology was clearly able to operate autonomously but the law raised the threshold to protect the jobs.

We can see this today. A diagnosis is only valid if a licensed clinician endorses it. Company accounts are only valid when an auditor signs them. In many settings, regulation and professional liability ensure that even a perfectly written contract is not valid until a lawyer approves it. Regulation requires humans to be the "legal provider" of the work. These laws will protect many jobs long after the machine is able to provide a flawless answer.

Because these decisions have substantial legal and societal implications, organizations and countries will make decisions that affect the height of the autonomy threshold. They will not passively let technology play out. The professional associations for doctors, lawyers, and engineers will press governments to forbid autonomous AI in their fields. Sometimes the decisive constraint is not the law, but insurance. Insurers decide what they underwrite and that may set the autonomy threshold.

This will be the defining political fight of our time. Tech firms and consumers will press to make autonomous deployment possible and reduce the cost of services, such as medical care. Professional associations will fight to prevent generative AI from autonomously doing their jobs and tasks. The results of these fights will vary by profession.

If productivity gains are sufficiently large, then it is unlikely that regulatory pressure will succeed at keeping the improvements at bay for long. The end of the elevator operators is instructive. A huge strike in New York in 1945, involving sixteen thousand operators, forced residents to climb dozens of floors. The strike effectively paralyzed the garment industry. It made operators expensive and unreliable, which played an important role in automating elevators.[39]

How regulation is deployed in this fight may be the largest source of variation in productivity growth between countries over the next decades. A country that locks its professions behind walls of mandatory human involvement, regardless of whether the human adds value, will risk repeating the mistake of the Ming China emperor, who in 1433 burned his ocean-going fleet and banned foreign voyages to protect his court from disruption. He also guaranteed that China would fall behind for centuries.

Preferences

Consumer preferences can prevent AI from operating autonomously even when the technology is able to do so. The same technical capability places different tasks on opposite sides of the autonomy threshold, depending on what the market requires.

In some domains, a 90 percent correct answer is a success. Consider an algorithmic recommendation by Spotify or YouTube. If the song or podcast is terrible, the consumer can click "next" and ignore it. No one is going to sue Spotify. Because the cost of making a mistake is low, AI can do the work of providing recommendations autonomously, without human review.

Now consider calculating the doses of a chemotherapy treatment. If the AI makes a numerical mistake and doubles the dose, the patient dies. The cost is unacceptable. Even a 99.9 percent probability of success is not good enough. A small margin of error puts human lives at risk and can lead to large lawsuits. A human must be in the loop and verify the decision.

This logic also shows up in robotics. During the 2026 Chinese New Year Gala, humanoid robots performed a dancing routine before an audience of hundreds of millions. This clever demonstration reflected clear strategic thinking. The engineers knew that if a dancing robot stumbled, the audience would just laugh. The cost of error was small. A reasonable rate of accuracy was more than enough for the dancing robot.[40]

Now consider the same robot holding a wine glass in a restaurant full of customers. The technical challenge may not be much harder than robot dancing. But the cost of error is significantly higher. Even if the robot drops just one glass out of every one hundred attempts, a busy restaurant will end up with a lot of wine-splattered customers and broken glass, not to mention potential lawsuits.

These examples illustrate the importance of considering the market's tolerance for failure. If the cost of technological error is high, the autonomy threshold also moves up.

The tolerance for errors is not the only way demand shapes the autonomy threshold. In some markets, customers value products and services with human origin. A live concert is preferred because the audience highly values the experience of sharing a room with a human performer. In these cases, the autonomy threshold is raised not by the risk of failure but by a positive preference for human provenance. We take up these demand-side determinants of the autonomy threshold in chapter 8.

The technology will undoubtedly continue its rapid progress. Verifiability, sycophancy, and continuous learning may be solved in the future. But even in the long term, autonomy may be constrained by laws and politics, and by consumer demand for humanness.

Jobs and earnings: the effect of automating cognitive tasks

To understand market value, we need to consider not only the direct effect of technology but also the impact of supply and demand responses. When AI crosses the autonomy threshold, the direct effect is substitution on that task. Then everything turns on demand. If demand is inelastic (when consumers are close to satiation), then human employment shrinks. If demand is elastic, due to the Jevons effect, human employment can persist, but only when humans remain complementary to the expanded output.

Three early examples of research pertained to simple coding tasks, customer service support, and writing tasks. In each case, the academic literature found that AI acted as a complement to human workers rather than as a substitute. Workers who were paired with AI tools became faster and, in some cases, produced higher-quality output. The gains were largest for the least skilled. Novices using AI closed much of the gap between them and experienced workers.[41]

This led to a discussion about whether AI would be broadly complementary to workers, raising productivity without threatening employment. If the pattern held, AI would look like a benign tool: Everyone would do better work, the weakest would improve the most, and the labor market would adjust smoothly. But recent developments in agentic AI suggest that most single-task, non-messy jobs are likely to fall below the autonomy threshold sooner or later, unless regulation protects them.

Consider the work of junior coders. On February 27, 2026, Jack Dorsey, cofounder of Block (the fintech company behind Square, Cash App, and Afterpay), posted his letter to shareholders explaining why the company was cutting its staff by 40 percent despite its strong performance. "A significantly smaller team, using the tools we're building, can do more and do it better. And intelligence tool capabilities are compounding faster every week," he explained. The market endorsed his explanation. Block's shares soared by 24 percent after his announcement.[42] Recently, companies like Amazon, Meta, Microsoft, and Verizon have all made sweeping cuts ostensibly for a similar reason.[43] Many tech jobs are highly specialized, consisting of a single task. Once technology crosses the autonomy threshold for that task, the supply will be provided by machines and the earnings of the job holders will collapse. Recent evidence by Crane and Soto (2026) shows this systematically. Using US labor-survey data, they show that five hundred thousand fewer coders are employed today than pre-LLM trends suggested.[44]

Next consider what happens with writing in the legal industry. In February 2023, Allen & Overy, one of the top-five "Magic Circle" law firms, became the first in the industry to deploy generative AI throughout the firm. Harvey, a specialized LLM for lawyers, was available to thirty-five hundred lawyers in forty-three offices. They used it to draft contracts, summarize due diligence documents, and research case law. The head of the firm's innovation group, David Wakeling, called it a "game changer," but he added a caveat: "You must validate everything coming out of the system. You have to check everything."[45] Harvey now serves more than seven hundred law firms and corporate legal departments across fifty-eight countries, including 50 percent of the top American law firms. Harvey was valued at $8 billion after its December 2025 funding round.[46] It drafts and summarizes contracts and briefs, and it does excellent research. Most of these tasks now fall below the autonomy threshold. If you give a model a term sheet and ask for a first draft, you will get something usable. The correctness of this type of work is externally verifiable. For transactional drafting, the autonomy threshold is low, and AI is crossing it.[47]

The market implications are rather bleak in all these cases. When AI crosses the autonomy threshold, it completes the entire task and produces outcomes without human intervention. A person is no longer necessary for the work itself, but a human is still needed for setting objectives and handling exceptions. Effective supply becomes enormous, and therefore wages for that activity drop.

For people working in jobs that consist of clean, simple tasks, only two factors will protect them: consumer demand for authenticity (which we discuss in chapter 8) and regulation. In the case of lawyers, for example, licensing and liability requirements will protect them. We can anticipate that lawyers will organize and put pressure on the legal system to ensure their physical presence continues to be needed. We are not convinced that such efforts will protect the drafting aspect of the job. If "a lawyer must sign" becomes a rubber-stamp task, then one lawyer can supervise far more AI output than one lawyer could produce alone.

Consider also the proliferation of direct-to-consumer telehealth services, such as Hims & Hers, that provide erectile dysfunction or ADHD prescriptions.[48] Yes, the law says there must be a human signature, but it is easy to find a signing doctor who can endorse, in seconds, the AI-written prescription.

As technology advances, the autonomy threshold will not be static. Jobs currently protected by legal or social barriers may soon be threatened by AI as the gap between human and machine performance shrinks.

Consider self-driving taxis. In early 2024, conventional taxi and ride-hailing drivers in Wuhan, a city in central China, organized a rare, large-scale protest against the rollout of Baidu's robotaxi service, Apollo Go. Videos that briefly circulated on Chinese social media showed that some protesters held banners demanding "fair competition," while others complained about being undercut by cars that needed no salaries and no sleep. Representatives of the protesters eventually obtained meetings with local officials who promised to study their grievances and to ensure the "orderly" development of autonomous vehicles.[49] But the officials did not promise withdrawal. Apollo Go continues to offer a five-mile ride for fifty US cents. To the extent that autonomous vehicle companies address safety concerns, many consumers will happily choose robotaxis over traditional services. The legal and political barriers are real, but they stand against a technological gap in favor of autonomous vehicles that grows every year.

Of course, none of the above means there will be no jobs for lawyers or programmers in the future economy. Much of what both professions do, as will become clear in the rest of this book, falls along the messier end of the spectrum. These are far from clean, single-task jobs. They are often, in fact, quite messy bundles of tasks.

CHAPTER 3

Tier Two: What Holds Jobs Together?

Strong and Weak Bundles

In 2016, Geoffrey Hinton, one of the founders of deep learning research, famously declared that radiologists should stop training immediately because AI would soon outperform them at reading scans.[50] His logic seemed solid because radiology involves recognizing patterns in digital images and AI is essentially a pattern recognition machine. Huge amounts of data exist to train the models.

The prediction was, at the very least, premature. It is true that many AI tools can analyze scans with precision. By November 2025, the FDA had approved more than 725 radiology AI models, representing about 76 percent of all approved medical AI devices.[51] Nevertheless, US radiology residency openings hit a record high in the same year, and average radiologist compensation rose to approximately $520,000, becoming the third-highest-paid specialty.[52]

Hinton made the mistake of confusing the task and the job. True, a radiologist reads scans. But she also trains other radiologists, triages cases, reviews the difficult scans, decides on treatments, and coordinates with other physicians on treatment plans. She also signs the diagnosis, making it legal.

This chapter explains why Hinton's mistake matters. AI can automate a task inside a job without automating the entire job. And most jobs are not limited to single tasks. They are bundles of tasks that must be performed together.

Not all bundles are the same. Some are weak. In those cases, the tasks within a job can be easily split apart, reorganized, and handled by AI. Other bundles are strong. If tasks in a strong bundle are separated, value is destroyed. That outcome occurs because the pieces rely heavily on shared context, responsibility, or coordination.

Understanding whether a job is a strong bundle or a weak bundle is the key to predicting whether AI will merely reshape a job or eliminate it.[53]

The patient does not buy a diagnosis. She buys a diagnosis that she trusts enough to act on and that has legal validity for treatment and insurance. AI alone can sometimes produce a diagnosis, but it cannot produce the bundle of services provided by the doctor. Even though AI can automate many tasks, that does not amount to automating the job. In many cases, the use of AI merely reshapes the job by stripping out some tasks and elevating others. To understand why, we must look at why tasks get bundled into jobs in the first place.

Bundled tasks and the risk of weak links

Consider a simple project: updating the pricing page on a company's website. Three experts are involved. Mark writes very well, Sarah is a great programmer, and Dave is an excellent lawyer.

Mark writes the new text for the website in a Word document. He sends it to Sarah, but she is busy so she only opens the email two days later. She realizes the text is too long for the design. She sends a Slack message to Mark. A few hours later, he edits it down and sends it back to Sarah, who writes the code. Then they send it to Dave for approval. He realizes the way the new price list is explained violates a consumer protection law, so he starts the process again. Two weeks and a few dozen emails after Mark wrote the initial text, the project is done. The total workload of the project was three hours, but coordinating three workers consumed two weeks.

Now consider a different organization. The firm uses a generalist, Jenny. She is a decent writer and programmer, and she knows just enough about the law to avoid going to jail. She opens the web editor. She changes the text. She revises it and presses "publish." Total work time: three hours. The output is not outstanding, but it is fine. Jenny is not better than Mark, Sarah, and Dave at each task, but she is faster because she can complete the bundle while eliminating the coordination costs entirely. She cannot misinterpret her own instructions, nor wait to answer her own questions.

Economists Gary Becker and Kevin Murphy argued that we do not specialize more because it costs too much to coordinate the work of many specialists. In markets and firms, people specialize until the cost of coordinating specialists becomes excessive.

> Most pediatricians in a city, or even in a single health maintenance organization, do not specialize in particular childhood diseases. No doubt they would learn more about a disease through specialization, but the additional knowledge would require greater expenses in coordinating their care with

> that of other pediatricians. For parents often do not know what is wrong with their children, and would need to see several pediatricians to get adequate care if each were highly specialized.[54]

Every time you split work across more people, you pay the coordination cost in terms of errors, meetings, and management. That's why companies bundle tasks within jobs to avoid paying this cost. However, this approach comes with risks and disadvantages. When different tasks are forced together into a single job for one person, the worker's performance is limited by his competence in relation to each task. A lack of competence in one task will worsen the overall results.

The best way to understand this is the O-ring theory, named after the space shuttle Challenger disaster in 1986. The Challenger was an engineering marvel that cost billions of dollars and had millions of parts. During a launch on an unusually cold morning in Florida, a simple rubber seal, an O-ring, froze, became brittle, and failed. The malfunction of this cheap part destroyed the shuttle and killed the astronauts.[55]

Economist Michael Kremer used this tragedy to explain an important economic phenomenon: In complex processes, the quality of the results is determined by the weakest link. This is crucial for economic development and growth. For instance, a country like South Africa with high human capital and lots of natural resources will not grow if the weak link of extremely unreliable energy keeps deterring companies from investing there.[56]

The same is true for individual jobs comprising a bundle of tasks. Imagine a highly specialized pediatrician who is a brilliant diagnostician, but who also needs to write notes in the electronic medical records. It turns out that she is a slow typist. Because the tasks are bundled, the hospital pays the doctor several hundred dollars an hour to enter data at the speed of a novice typist. Her slow typing speed is a bottleneck in the bundle. One weak skill degrades the value of everything else and increases costs. If the written record has a typo that leads to the wrong medicine being prescribed, the outcome is even worse.

How AI changes bundles

To coordinate with another human, you need empathy, clear communication, a shared context, and similar schedules. You cannot delegate a task to someone who is sleeping or busy. If it takes a long time to explain a task, it is often better to do it yourself.

AI has none of these restrictions. It is a subordinate that is available twenty-four hours a day. It has no ego and no conflicting priorities. You do not need to be polite. There are no feelings to manage. There is no waiting for a coworker to be available.

Let's use the pediatrician who wrote her own notes as an example. She might benefit by hiring a human scribe, but she might exclude that option because a new employee would introduce too much friction in the form of a large coordination tax: hiring the scribe, training him, communicating with him, scheduling work, and holding meetings.

Today, the scribe is an AI that listens to the meeting, extracts the relevant data, and writes the note instantly. The doctor checks it on her screen. The coordination cost is much lower. The "documentation" task has been unbundled from the "medicine" task. AI acts as a perfect scribe. This allows the bundle to break. The doctor continues to provide the diagnosis, which is her main task. She unloads the documentation work, which is her weak-link task, to the machine.

A recent experiment at Procter & Gamble, led by Raffaella Sadun and her coauthors at Harvard Business School, makes this point directly. They compared three configurations: a single specialist working alone, two human specialists collaborating, and a single specialist working with AI. The single specialist with AI improved her output by 0.37 standard deviations. Adding a second human specialist improved output by only 0.24 standard deviations. The two-person team had more combined expertise, but the coordination tax of communicating different contexts and perspectives ate into the knowledge advantage. The human-AI pair dominated both alternatives, not because the AI knew more than the second specialist, but because the coordination cost was lower.[57]

Coordinating with AI is less expensive than with a human, but it is not free. The relevant question is whether the gains from separating a task and handing it to AI exceed the losses from breaking the bundle. That is why we need to distinguish between weak bundles, which can be split apart at low cost, and strong bundles, where separation destroys value. Three conditions make unbundling costly. When any of these conditions hold, the bundle survives. Otherwise, the bundle is weak and comes apart under AI.

Unpredictable demand. The first condition comes from the demand side. Within a job, some tasks are better suited to AI and others to humans. In principle, you can assign the tasks accordingly. But that is often difficult in practice because the demands of customers are complex, and you do not know which tasks the customer will need.

One of us (Yanhui) spent time in Hangzhou with a leading customer service company that had pioneered the use of a domain-specific LLM in high-end service. After two years of deployment, human agents still played an indispensable role.

At this company, human agents handle a broad mix of work: answering inquiries, resolving complaints, promoting products, collecting feedback, and managing customer relationships. Their performance is evaluated along six dimensions: accuracy, speed, completeness, flexibility, sales awareness, and customer relationships.

On the first four, AI already performs extremely well. The last two remain much harder to automate. Sales awareness involves recognizing what the customer is not explicitly saying. If a human agent senses that a customer is on the verge of leaving, she must quickly steer him toward an alternative product. Customer relationship management requires a different kind of judgment: knowing how to respond without creating frustration. News about an interaction that offended a customer can spread quickly on social media and become very costly for the firm.

The division of labor, in principle, is clear. AI is better at routine parts of the interaction and humans are better at the delicate ones. The difficulty is that customers do not arrive in clearly sorted categories. At the start of an interaction, the firm often does not know whether the interaction will remain a routine inquiry or if it will become a retention issue. As one senior manager put it, "Such unbundling will require frequent switches between the AI and human modes. The coordination cost would be too high."

Now contrast this with a case in which demand is predictable.

Strategy scholar Nan Jia and coauthors studied a large telemarketing company selling credit cards by making outbound calls. The sales process had two clear stages. First was lead generation, which involved calling customers to gauge interest. This was a scripted and repetitive task. If the customer was interested, the second stage sought to convert the interest into a sale, which involved a far less structured conversation.[58]

The company assigned AI to the first stage. Then a human agent entered the conversation. This worked because the sequence of demand was known in advance. Stage one was always about identifying interest. Stage two was always about closing the sale. There was no ambiguity about which capability was needed or when. The handoff occurred only once, at a clearly defined point.

That is a weak bundle. When demand arrives in a predictable sequence, the work can be partitioned cleanly. But when customers arrive with mixed and shifting needs, and the nature of the interaction is only revealed along the way, then unbundling creates costly problems like misrouting or constant switching.

Production spillovers. The second condition comes from the supply side. Sometimes tasks belong together because doing one makes you better at the other one. When the spillover is large, separating the tasks reduces the quality of the whole job.

Radiology is a good example. Reading a scan requires more than pattern recognition. A radiologist who has spoken with the patient and reviewed the clinical record can do a better job reading the mammogram than one who sees only the image. She knows whether an anomaly is the result of a previous condition. If interacting with the patient is taken out of the bundle, the diagnosis itself may get worse.

The same is true in law. By the time a litigator cross-examines a witness, she has usually spent weeks reading discovery, taking depositions, and drafting briefs. Those tasks help build the case in her mind. They teach her where the weak points are, which details matter, and which story the other side is trying to tell. The lawyer who only relies on AI may walk into court without knowing the case deeply, and the cross-examination will suffer.

This is the logic of production spillovers. One task creates knowledge or instinct that improves another. When that is true, the tasks form a strong bundle.

Measurement challenges. The third condition is about measurement. When AI produces one piece of the work and a human produces the rest, it is hard to know whether mistakes in the final product were caused by the human or the AI. If the software crashes, did the AI-generated code cause the problem or should blame be place on the engineer who stitched it together?

Economists Armen Alchian and Harold Demsetz call this issue the "metering problem." When several inputs jointly produce an output, their distinct contributions are difficult to disentangle. The standard solution is to bundle the tasks into one person's job. Rather than trying to assign blame task by task, you make that person responsible for the whole thing.[59]

This is why a senior engineer can rely heavily on AI and still be fully accountable for the project. If the system fails, she answers for it, not the model. If the bundle is broken so that AI writes one module, a second engineer integrates it with the other modules, and a third one tests everything, then assigning blame for mistakes becomes very difficult.

The managers in the Hangzhou customer service company

made the same point from a different angle. They kept coming back to verifiability, liability, and responsibility. In other words, if a bot offends a VIP client, or if an AI-drafted response triggers a regulatory inquiry, then the algorithm cannot be fired or sued. Somebody must own the outcome. That somebody must see the whole bundle.

In summary, when the qualities of the parts cannot be judged on their own, the measurement challenge creates a strong bundle. Companies ensure that someone is held responsible by bundling the tasks within one person's job.

The three conditions described above offer a test to assess whether a job can survive AI. First, how predictable is the demand? Can you forecast which tasks the demand will require? Second, how large are the production spillovers? And third, how divisible is the measurement? Can you verify the quality of each task independently, or can you only judge the final product as a whole? If the demand is unpredictable, or if the spillover is large, or if measuring the task separately is difficult, then the job will survive. Otherwise, AI will unbundle tasks out of the job.

What do surviving bundles look like?

Not all residual human work is equal. A consistent pattern runs through decades of data: The labor market increasingly rewards people who combine analytical thinking with the ability to work well with other people.

The economist David Deming documented this pattern across American occupations from 1980 to 2012.[60] He classified jobs according to their cognitive skill and social skill demands. He found that jobs high on both dimensions grew by roughly twelve percentage points over three decades. Jobs that were cognitively demanding but low on social interaction (number crunching) shrank by about three percentage points. The demand increase was not for "soft skills" but for people who could both think and coordinate.

Catherine Weinberger found the same pattern from the wage side.[61] Comparing cohorts of American workers, she showed that the relative wages of those with high cognitive ability rose mainly for those who also had strong social skills. Again, the extra payoff from having *both* grew stronger with each cohort. Social skills were not a substitute for analytical ability, but they were a multiplier of it.

Our three conditions for strong bundles (demand, production spillovers, and measurement) help shed light on these findings. Jobs that combine difficult judgment with difficult relational coordination tend to score high on all three conditions. The *demand,* as explained earlier, is unpredictable. This means that workers often face complex problems that can only be understood during interactions with customers and coworkers. The *production spillovers* are strong. As a result, understanding the problem helps the worker explain it, and explaining it sharpens her understanding. And the *measurement* is indivisible; that is, the value of the work cannot be decomposed into "the analysis part" and "the relationship part" because they are jointly produced.

This is why our "strong bundles" argument is more precise than just saying "soft skills matter." The greeter at the store and the receptionist at the desk perform social interactions, but if their cognitive demands are low, AI and self-service kiosks can replace them. What the data show is that the interaction between social and cognitive capability produces the strongest employment growth and wage growth.[62]

In our view, jobs that combine difficult judgment with difficult coordination will most likely be strong bundles. They are not protected because they are vague or because they involve "people skills" in some generic sense. They are protected because thinking and relating are tightly linked in the work itself.

Jobs and earnings: the effects of unbundling

The labor market implications for a multitask job are different from those derived from a single-task job. When a job consists of one task and AI crosses the autonomy threshold for that task, the replacement is clean. But jobs usually comprise bundles of tasks. The threshold for replacing the whole job is much higher than the threshold for automating any one task inside it. Even if AI can perform one component well, someone may still need to hold the bundle together, take responsibility for the outcome, and bear the coordination costs that will arise if the work is split across multiple agents. This is where Hinton and other tech people went wrong about radiologists. They confused a task with a job.

Thinking in terms of bundles also helps explain a puzzle in the data. Occupations that are viewed as highly exposed to generative AI have not necessarily experienced a significant decline in job postings. One of us (Yanhui) has studied twenty million job postings from China's five major recruitment platforms between August 2024 and June 2025. He analyzed the impact of the sudden arrival of DeepSeek V3 in December 2024. Consistent with what we observed in the customer-service company in Hangzhou, the number of job postings for customer service agents in China barely changed after the entry of DeepSeek, although more jobs required the skill of using generative AI.

The market prices the bundle, not single tasks. The distinction between strong and weak bundles, therefore, determines what happens to earnings.

In a *strong* bundle, the direct technological effect of AI is positive. AI raises productivity on the tasks it can handle while the human retains the full job. The equilibrium effect tends to reinforce the direct effect. The market still buys the same bundle of judgment, coordination, and responsibility; however, with AI the worker can deliver it all more efficiently. Because the current supply of qualified workers in certain fields remains constrained (radiologists still require more than a decade of training and skilled customer service agents still require deep domain knowledge), earnings often remain unchanged or even increase.

AI's effects on *weak* bundles are very different. The technology strips away the tasks that used to be bundled with the core. The direct technological effect is ambiguous. The worker loses some tasks to AI, which reduces the scope of the role. The remaining core tasks are still viable for humans, and may even become more productive with AI assistance, but the job becomes narrower and simpler.

The thinning of the job has a negative equilibrium effect. When the peripheral tasks are eliminated, the job becomes less complex. This means there is a larger pool of people capable of performing it. Workers whose employment was protected by their ability to handle the full bundle will now face increased competition. The resulting increase in labor supply pushes earnings down, even if demand for the core tasks remains stable. The lower barrier to entry leads to a wage decline.

This is why the distinction between strong and weak bundles is so consequential. In a strong bundle, AI raises productivity while the coordination cost keeps the job intact and the supply of qualified workers remains constrained. In these cases, earnings hold steady or rise. In a weak bundle, AI raises productivity on the surviving tasks but lowers the complexity of the overall role. When that occurs, the supply of eligible workers expands and the earnings of workers go down.

There is one important caveat. The equilibrium effect of unbundling can be positive if the use of AI creates sufficient new demand. For example, many promising software projects never start because companies cannot justify the engineering costs of developing new programs. However, these projects become viable when AI coding tools substantially reduce the cost of software development. This is the Jevons effect at work: Cheaper code creates enormous new demand for software that was previously too expensive to build. Perhaps that is why the US Bureau of Labor Statistics (BLS) optimistically projected employment for software developers to grow 15 percent from 2024 to 2034, five times the average for all occupations, with about 129,000 openings per year.[63]

For the reader thinking about career choices, the implication is quite clear. The question is not just "Which of my tasks can AI do?" but "What does my job look like after the bundle absorbs tasks from adjacent roles?" The answer, across a wide range of occupations, is that the post-AI bundle requires more social coordination, more judgment, and more ability to adapt to new roles. This is uncomfortable for specialists who built their identity around a narrow skill. But it is encouraging for people who can operate across boundaries, for those who can analyze a problem and explain it to a skeptical audience, or who can manage a technical system and the organizational politics of deploying it.

Once AI enters the bundle, the internal structure of the job changes. Sometimes AI removes the routine tasks beneath the human. Sometimes it delivers expert knowledge downward to less-skilled workers.

The next two chapters examine these two possibilities. In chapter 4, we study what happens when AI sits *below* the human in the hierarchy and automates routine work. In chapter 5, we study the opposite case, when AI sits *above* the human and delivers expert knowledge downward.

CHAPTER 4

The One-Person Company

Robots Below

What happens when AI takes over the routine layer beneath a skilled human? It lets one capable person leverage her judgment across many routine tasks and sometimes across an entire firm.

In June 2025, the AI app development company Base44 was sold for $80 million. The firm's technology allowed users with no programming knowledge to create functioning applications by using natural language. Notably, Base44 had no employees. Founder Maor Shlomo built the firm and managed it on his own.[64]

Shlomo was not superhuman. He was a competent engineer who used AI to generate the interface, write the marketing texts, provide customer support, and write and revise his contracts. The AI-based functions were not as good as what a specialist team could have delivered, but they were good enough. Because Shlomo did not need to coordinate with anyone, he could move so fast that traditional competitors could not catch up. Six months after launch, Base44 had hundreds of thousands of users. The acquisition valued the firm at a price that would have seemed absurd for any one-person company in the past.

This is a model for a one-person company. A founder with strong core skills uses AI to leverage his expertise, patch up his weak links, eliminate excess coordination, and build a firm without a team. AI lets a single competent person quickly build a complete bundle of tasks while rivals struggle with the burdens of coordination. The one-person company with AI can do work that was previously done by an entire hierarchy: supervisors giving commands to subordinates and cross-functional specialists coordinating efforts to solve problems.

This is a classic example of AI crossing the autonomy threshold, at least in relation to certain tasks. AI can do the routine work and handle problems with known solutions. But in this tier-two example, AI remains *below* the threshold. A bundle of interdependent tasks remains above the autonomy threshold. Thus, the work still requires a human expert's judgment.

In this chapter, we ask what happens inside the bundle when AI takes over part of the knowledge hierarchy from below.

Management by exception

Why does the hierarchy exist in the first place? One of us (Luis) has spent the better part of his career studying how organizations utilize the knowledge of experts, whose time is scarce. Common, low-stakes cases are handled by many workers with basic knowledge at the bottom of the hierarchy. The high-stakes problems, the exceptions, are funneled up to fewer, more knowledgeable experts at the top. This is "management by exception."[65]

You see this organizational structure everywhere. The hierarchy is built around who has the highest level of knowledge. In a law firm, for example, a paralegal escalates unusual issues to an associate, who escalates those matters to a partner. In a hospital, a nurse escalates to a resident, who escalates to an attending physician. In a manufacturing plant, a coordinator escalates to an engineer, who escalates to the head of engineering. The organization is a pipeline of exceptions moving upward. The directions (solutions) move down. Those with less knowledge are directed, or helped, by those with more. Such an organizational structure is constrained by two factors: the cost of acquiring knowledge, in terms of learning and training, and the cost of transferring knowledge, in terms of time.

AI changes this structure radically because it makes knowledge abundant. And, unlike human intelligence, AI has no time constraint. A human expert's knowledge is constrained by time, whereas an AI system can be replicated across vast amounts of computing power and applied simultaneously to millions of problems. That means it can be used at scale.

AI allows a person like Shlomo to take care of the routine tasks cheaply and quickly. This is what we call "robots below." Depending on the type of industry, AI tools can do the support work that used to require junior staff, specialist departments, or outside contractors. A human stays on top of the knowledge hierarchy to deal with the exceptions: decisions that require human judgment, taste, accountability, and contact with the messy outside world. Those exceptions flow upward to the expert, who becomes more valuable because her knowledge is now leveraged by a very large number of AI agents.

The founder's dilemma

A new business is a chain of synchronized functions. You need a product that works, but you also need pricing that makes sense, a message that people understand, a way for them to find you, a way for them to pay you, and a way to support them when something breaks. You need contracts and compliance that keep you out of trouble. You need basic analytics to know what is happening. You need designs that signal competence rather than desperation.

Most founders are strong in one or two of these functions and weak in others. A technical founder may be excellent at building the product and average at everything that surrounds it. In the early months, those surrounding pieces are essential. Without them the business fails.

This is a startup's version of O-ring logic. The traditional solution is to hire specialists or to find cofounders. The startup brings in a marketing expert, a financial director, and a designer. But this creates new problems, such as the cost of paying specialists. Finding them and managing them takes a lot of time and scarce attention. In early-stage businesses, there is not enough work in any function to justify a full-time specialist. She may need ten hours of legal work one month and seventy hours the following month.

So the founder is faced with a dilemma. Generalists have weak links, but specialists need costly coordination and cannot work outside their realm of expertise, at least not in accordance with the varying needs of the startup. The founder faces a choice between spending less but being limited by her own capabilities and spending more to hire and manage specialists.

AI makes smaller firms viable

Now provide a startup founder with access to AI. The text of the landing page that would have taken hours to write with considerable effort, and that might have produced mediocre results, can be written in minutes and can then be refined. The terms of service that would have required legal advice or potentially dangerous shortcuts can be generated from a simple prompt and then revised to detect obvious mistakes.

AI-generated results will not be as good as what a top specialist could do, but AI only needs to be good enough to eliminate the weak link, as occurred for Base44. A fast-moving startup in a new business environment is much more valuable than a slow mover waiting for specialists to repair weak links.

These changes, enabled by AI, make small firms more viable. The tasks that can be done by AI require neither employees nor outside contractors. Thus, the activities that were previously grouped inside a business can be done by a single person. The minimum efficient scale for a serious business has dramatically dropped. Activities that used to require teams now require individuals.

Nevertheless, the quality of the entrepreneur using AI still matters. The entrepreneur's judgment remains essential. Knowing what to ask and choosing the problems to deal with is a key skill that only a human can do.

A study of 640 Kenyan entrepreneurs who received a GPT-4-based mentor delivered via WhatsApp found that there was no average effect on performance. There was, however, huge heterogeneity between different entrepreneurs. High-performing entrepreneurs improved their profits by about 20 percent with AI advice, but low-performing entrepreneurs saw their performance decline by about 10 percent. Knowing what to ask was a critical factor in being able to use the AI advice. The distinction between the high-performing and low-performing groups appeared when the topics in the study became more difficult and challenging. Because they lacked baseline knowledge, the low-performing entrepreneurs tended to ask AI for answers about harder, more open-ended problems. AI advice was less reliable for those types of questions. By comparison, high-performing entrepreneurs typically asked bounded, operational questions that AI usually handles well.[66]

Specific skills were also relevant in the study's results. The better entrepreneurs were good at structuring problems; that is, knowing how to break a messy situation into bounded questions that AI can answer well rather than throwing open-ended dilemmas at it. The job of the better expert, according to management-by-exception logic, is to restructure complex problems into tractable subproblems, which is exactly what the high-performing Kenyan entrepreneurs did.

In summary, AI cheapens cognition while raising the premium on judgment and taste. Given clear, specific instructions, AI can implement plans faithfully and produce results worthy of praise. But the guidance of the entrepreneur, and her ability to run a messy business, remain critical.

On the production side of the founder's job, AI empowers her to carry out work that used to require an entire team. However, AI's effect on the demand side is different. AI cannot tell you whether anyone wants what the entrepreneur is producing. Figuring that out requires the ability to notice what is missing before anyone asks for it.

Finding unmet needs

The world is full of unmet needs, but they are fragmented and heterogeneous. Millions of small niches of consumers want slightly different things. The entrepreneur's job is to find out which of those niches she can serve. She must figure out what consumers want, which requires research and discovery; and she must assemble a bundle of tasks that satisfies that demand. The tasks might involve providing content, products, experiences, engagement, and community. The whole job requires more than a one-time act of "idea generation." It is continuous: reading the audience, adapting, experimenting, and maintaining attention and social capital over time.

Consider Kat Norton (better known online as "Miss Excel"), a senior consultant at Protiviti in early 2020. She would fly out of New York City most Mondays to do bank reviews and then fly home to write them up. She would use Excel as part of her daily work, and she got good enough to build training workshops for coworkers.[67]

Covid-19 suddenly stopped her travel. She found herself back in her childhood bedroom, wondering what to do with her life. Not one to quit, she repackaged her in-person Excel training material into fourteen-second TikTok videos with music. She started posting in June 2020. By her fourth video, she was already getting huge view counts. The same explanation that used to have an audience of one now reached thousands. Seeing the success, she began posting her content across platforms, including a YouTube channel for longer-form training videos.

In early 2021 she left Protiviti to run Miss Excel full time, selling her "Excelerator" training as an online course. She told

Business Insider that she had already made $80,000 from classes that year. A course-platform case study reported that her first six-figure month was April 2021 (more than $105,000). Today, she has built an audience that tops a million on TikTok and Instagram and she has maintained a YouTube channel with hundreds of videos.

Norton's experience illustrates both halves of the founder's job. AI can help with the production side: scripts, captions, course materials, and customer replies, but AI did not make her successful. She spotted a niche need that nobody was serving well. Seeing that people were frustrated by spreadsheets, she figured out that they wanted fourteen-second TikTok posts with music, not hour-long webinars. She kept showing up, building an audience that trusted her.

Good entrepreneurs who have skills to discover and build demand are likely to remain valuable.

Jobs and earnings: niches and superstars

The direct effect of AI is that it empowers individuals to produce more, faster, and at higher quality than before. But we still face the equilibrium question: What happens when everybody has the same tools?

When everyone can ship a product, write content, and launch a course, there is almost no barrier to becoming an entrepreneur. The bottleneck comes from the demand side. Consumers have fixed time and limited attention. As the number of entrants competing for attention increases, average earnings compress. If anyone can try to start a business, and trying is cheap, the expected payoff from trying cannot be large for most people. Two types of markets are likely to benefit, depending on demand differentiation: superstar and niche markets.

Superstar markets

As the cost of starting a business drops, more solo founders will be able to compete; however, the same cost reduction can also amplify the best founders' returns. When production is cheap and distribution is global, the difference between the best and the second-best can be enormous.

Think of a carpenter. He buys wood and takes a day to make a chair. To make more chairs, he needs more wood and another day. If he wants to make one million chairs, he needs one million workdays. Time limits his market and income. He has no leverage.

Contrast this with Taylor Swift. Each song she records will get played millions of times by her fans. She has enormous leverage. Before the existence of recording technology, a singer could reach at most a few thousand people per performance. Local talent would sing for local audiences. The best singer in Milan competed with the best singer in Naples for their respective audiences. Recording technology changed everything. The best singer in the world could compete with everyone for every listener. Technology relaxed the capacity constraints.

The distribution of talent is believed to be normal, appearing as a bell curve. So, one would expect the distribution of earnings to take a similar shape. Not so. There is a huge concentration of rewards at the top.[68] Leverage of talent is a key reason for this.[69] In the digital world, a product can be multiplied an infinite number of times without any additional cost. This mechanism permits the work of one person to scale across the world. In statistical parlance, the earnings follow a "power law." The top 1 percent captures a large share of the value because their work scales.[70]

AI amplifies the conditions that allow digital technology to provide large markets for talent. It increases the supply of talent, which makes the quality of talent matter more. In 2025, Meta offered top AI researchers compensation packages exceeding $100 million. Sam Altman, OpenAI's CEO, confirmed the staggering figures.[71] Benjamin Mann, an Anthropic cofounder, defended the economics of these enormous packages. "To pay individuals $100 million over a four-year package, that's actually pretty cheap compared to the value created for the business."[72]

Such compensation would have been unthinkable five years earlier. What changed? Not the quality of the researchers. Instead, AI's capacity to reach millions of users dramatically amplified the value of their knowledge. What changed was not talent, but leverage.

Niche markets

The market for AI researchers, in which a handful of people capture nearly all the value, is highly unusual. In many markets, consumers have different preferences and variety emerges to satisfy different tastes. These differences are often large enough that no single provider can satisfy them all. Low distribution cost and algorithmic matching allow niche producers to each find their own scattered audience.

When Luis visited Jin and Yanhui in Hong Kong, he enjoyed an Uber trip with a very entrepreneurial driver. Peter (not his real name) was running a business in the United States. He would buy household items in China, export them to an Amazon warehouse, and sell them under his own brand. For Peter, AI was a huge improvement. "It's a completely level playing field with the big companies," he said. "All the functions for which I'm not great—analyzing data, thinking about inventory levels, writing product copy, making pictures and logos, accounting, marketing, finance—can be done with the LLMs that I am using for free."

AI has dramatically improved Peter's ability to compete with bigger rivals. He is a living example of how to overcome weak links. His central competence is finding products and understanding Amazon. AI does everything else. He competes with businesses that are ten or one hundred times bigger, but he does not have to incur his competitors' coordination costs, and he gets good-enough advice from AI.

This AI capacity is not limited to commercial entrepreneurs. *Political* entrepreneurs discover unmet demand for policy or representation among fragmented constituencies. A local organizer who is trying to build a coalition around housing reform is also discovering demand. She identifies people with a shared grievance, figures out what they will mobilize for, and maintains their engagement through personal relationships and credibility. AI gives her better tools for polling, messaging, data analysis, and content production. But she retains the core job of understanding a community and earning its trust. Business entrepreneurs, social entrepreneurs, and political entrepreneurs all get more leverage from AI, but they still have to do the messy, relational work that defines the role.

Research on Spotify shows that its algorithm reduces search

costs and enables users to match unusual preferences with unusual producers across the world. This allows many "microstars" to make a good living by serving small, specific audiences.[73] AI helps these niche musicians to produce and distribute quality music at a scale they previously could not afford. But AI tools fall flat without the lessons learned by Miss Excel. She found a specific audience with a specific need and served it with sustained care and personality.

The squeezed local entrepreneurs

Artificial intelligence benefits superstars and niche producers, but it hurts the mid-tier producer who survives on a local advantage. Before AI, he was the best available solution within a geographic or social radius. AI reduces that advantage in two directions. *From above,* the global superstars become accessible to every buyer. There is no need to hear the local singer if Taylor Swift is available. *From below,* AI tools allow niche producers to find their niche. Meanwhile, the mid-tier producer can no longer rely on providing good products and services to local customers. He is defenseless. He is neither distinctive enough to meet the niche nor talented enough to dominate at scale.

Think of a local graphic designer. Ten years ago, she earned a decent living making menus for local cafés and logos for the local plumber. She was not the best in the world, but she was the best in the city. Now, the local café owner can hire a star freelancer from the other side of the world for the same price. Or she can use an AI tool, such as Midjourney or Gemini's Nano Banana, to generate a logo for free.

It is easy to feel jealous when one sees only the most salient cases and the big paychecks, or to overestimate one's chances of reaching the top of a field and to succumb to "selection bias." We only see the winners. We do not see the many people who tried to build a large audience and failed, even though they put in as much work as the winners.

The right response to this uncertainty and risk is not to avoid leverage, but to understand why it is there and how it plays in different fields. For some people with extreme talent in their field, the right strategy is to make a big bet and join the race to be in the top 0.1 percent of the distribution. For the rest of us, it is good to focus on a career or niche that includes leverage. In other words, we should not sell our time to only one client or product, like the carpenter in our example; rather, we should sell our products, skills, knowledge, and reputation in ways that scale.

This chapter examined what happens when AI sits *below* the human in the hierarchy of knowledge. The founder patches her weak links, the expert scales her reach, and the market rewards those who combine production leverage with the messy, relational work of demand discovery. The next chapter flips the ladder and looks at robots *above.* We describe how "expertise in a box" expands the scope of middle-skill workers, and what happens when that expansion translates into better jobs rather than a race to the bottom.

CHAPTER 5

The Rise of the Middle Class

Robots Above

What happens when AI delivers expert knowledge to less-skilled workers? It expands the scope of middle-skill workers. That does not necessarily mean these middle-skill workers are better off. They gain only when demand expands, barriers to entry remain, and the bundle holds together.

The night shift staff at the pediatric pulmonology wing of Hospital Ramón y Cajal in Madrid works as they do in many other hospitals. The ward nurses monitor oxygen levels and other vital signs of children with respiratory trouble. They follow a detailed protocol and deal with routine visits. At any sign of trouble, they call the pediatric pulmonologist.

Imagine workers on the same shift with a high-quality AI system. The nurse enters all the information as before, but now the AI system can deal with a larger share of apparent problems. Only in rare catastrophic situations must the doctor be called. The nurse must still do the physical work, monitor the children, calm the parents, and call security if needed. But she now has a specialist in her pocket. She can deal with more patients, reducing their wait times.

As "expertise in a box," AI can relax the constraints imposed by previously scarce cognition and expand the scope of middle-skill workers. The nurse practitioner is able to handle cases that previously required a doctor. Whether nurses' wages rise depends on what remains scarce once everyone has the tool.

In the previous chapter, we discussed what happens when AI automates the routine layer of the knowledge hierarchy and lets the experts scale (the "robots below" scenario). In these cases, the best software engineer benefits from running hundreds of instances of Claude Code.

In this chapter, we look at what happens in "robots-above" scenarios. AI provides expert knowledge to the upper "production" layer of the organizational hierarchy so that the nurse can solve problems that previously required escalation to a doctor. Both chapters are about tier two, which, as a reminder, pertains to bundled professional work where the autonomy threshold for the whole job remains high.

The end of skill-biased technological change?

In the IT revolution, computers improved the wages of the most-skilled workers. The gains were "skill biased." To use an Excel spreadsheet or a database or analytical software, a person needed to have high cognitive skills. Computers complemented workers who could program, analyze data, or manage information systems, but it left behind those who could not. As a result, the wage gap between skilled and unskilled workers widened.

Many economists are worried that AI will repeat this pattern.[74] Automation displaces workers whose jobs are heavily task-based, unless new tasks are created. This negative outcome could occur when we consider the impact of AI on simple, single-task jobs. But when we think of jobs as bundles of skills, there are good reasons to expect that AI will not cause widespread displacement.

The computer revolution rewarded human cognition because computers lacked cognitive skills. AI reverses the direction. When the machine supplies cognition, the scarce factors shift to what the machine cannot provide: physical presence, social trust, accountability, and emotional engagement. These are disproportionately bundled into middle-skill jobs. Many of the skills present in those jobs, such as physical, emotional, and social skills, are complements to AI. And in some cases, as with the nurse, they can be rebundled with cognition once AI makes expert cognition abundant.[75]

When AI favors the less-skilled

Consider the hierarchy between a doctor and a nurse. A "complete medical case," such as a hospital visit, involves two activities. The first is production (or execution), which is the physical work, the monitoring, the relationship with the patient. This is inevitably human work. The second is problem solving, which involves the diagnosis and the decision to escalate the case or refer a patient to a specialist.

Because expert knowledge is expensive, the nurse specializes in production and handles only a small sliver of the cognitive work.

The doctor adds the expert cognition. This form of specialization "deskills" the nurse, who must escalate any complex problem. Nurses often complain about feeling constrained in what they can do.

When AI makes expert knowledge abundant, the nurse can solve a larger share of problems without escalation. Her job is rebundled. The job holds together strongly (see chapter 3) because patient demand is unpredictable, the physical care generates knowledge that improves the cognitive response, and the quality of the outcome cannot be deconstructed into an AI task and a human task. Rather than specialize only in production, the nurse also takes charge of a larger share of the diagnostic work. The organization needs fewer expert hours per unit of output. The entry-level worker becomes more valuable.

The computer revolution gave the *expert* more leverage. By comparison, AI gives the *non-expert* more capability. The direction of the complementarity reverses; in other words, the bottleneck shifts from cognition to the physical, social, and emotional skills that the machine cannot supply. There is also a reduction in the share of cases that need expert time, hence a reduction in the demand for experts and in the wage premia they can command.

When machines cheapen the thinking tasks, the remaining scarce input is the part of the job that cannot be reduced to text. In a medical context, a human is needed to help the patient understand, accept, and act on an intervention. A medical professional must be accountable for the diagnosis and treatment plan. If the patient does not trust the advice, if the team does not coordinate, if the person in charge will not authorize the decision, then the medical intervention fails. These relational and physical tasks are, in many settings, the binding constraints.

This pattern appears across many settings. For example, AI significantly helps less-experienced taxi drivers. Being an effective taxi driver requires two distinct skill sets: operating the vehicle and solving the "cruising problem," which is knowing exactly where to drive to find the next fare. Veteran drivers intuitively know which neighborhoods peak at specific hours and which corners reliably produce passengers, but new drivers lack this localized knowledge. For novice drivers, this uncertainty leads to lower income and increased anxiety. Consequently, there is a higher turnover rate among inexperienced drivers, creating a retention problem.

To address this, a tech company introduced a machine-learning tool called AI Navi in December 2019. During a free trial, the tool, which delivers demand-optimized routes in real time, was provided to about 520 drivers in Yokohama, Japan, a city of 3.77 million people. Researchers found that drivers who used this tool reduced their passenger search time by 5 percent on average. More importantly, less-experienced drivers (those in the bottom third of prior performance) benefited most. They saw a 7.4 percent reduction in cruising time. These findings suggest that companies can, by using AI, complement the skills of the drivers and increase the earnings of their workforce.[76]

By making more knowledge accessible to workers, AI also expands the talent pool for firms. An example of this comes from KFC China, which has been rapidly increasing the number of stores it operates. To sustain the growth, KFC China requires each store to have a competent Restaurant General Manager (RGM). These RGMs must do everything. Their role involves bundled, messy, human-centric tasks, such as motivating a team or handling an angry customer. The RGMs must handle highly demanding analytical tasks, such as demand forecasting, labor scheduling, and inventory management. The problem is that not everyone is equipped to do everything well. Many managers possess excellent social and leadership skills, but they naturally struggle with the heavy cognitive and analytical demands of the job, which can often feel like "rocket science."

To deal with this problem, KFC introduced an AI toolkit designed to act as an operational "super brain." The AI acts as an assistant to the RGMs, giving them suggestions on how much inventory to stock and how many employees to schedule. The RGMs still have the final decision, but because the AI suggestions are usually accurate, they rubber-stamp them.[77]

The AI tool is particularly helpful to managers with fewer analytical skills. The RGMs used to spend lots of time on number-crunching tasks, often making errors. With AI, they now focus their energy on what they do best: applying social skills to lead teams and connect with customers. When their charm manages to win many new customers, they don't need to worry about whether they have ordered enough inventory to serve them.

The pattern across domains is the same. AI encodes and distributes expert knowledge, so those with less education and experience gain the most. The machine pushes knowledge downward.

However, these direct gains describe what happens to one worker. When *every worker* has the same tool, the question is no longer whether each worker becomes more productive (they do); rather, the question is whether the scarcity that sustains one worker's wages will survive when everyone has the tool. If the hard cognitive step becomes easy, then entry barriers can fall. More people can do the work, or adjacent roles can expand into it, raising competition. Whether wages improve depends on the equilibrium of the economy, on how the whole system adjusts.

Jobs and earnings: a new middle class?

Even if AI favors less-skilled workers by making them more productive, this does not mean that their wages will increase. The outcome depends on three conditions. First, demand must be sufficiently elastic so that cheaper, more abundant supply finds more buyers rather than causing the displacement of existing workers. If there is large unmet demand waiting to be unlocked by lower cost, the upskilled workers will have somewhere to work.

Second, entry barriers must remain high enough to prevent wage gains from being stifled by competition. If AI eliminates the cognitive barrier and the remaining tasks require little training, then new entrants will flood in and wages will fall even as productivity rises.

Third, the task bundle must hold together. Cognitive and noncognitive components of the job must remain inseparable in practice so that the upskilled worker captures the full value of the AI-augmented output. For example, if cognition is automated centrally while physical and relational tasks are commoditized separately, then individual workers will lose bargaining power. Consider three occupations: healthcare, the trades, and back-office clerical roles.

Healthcare. This professional sector meets all three conditions (mentioned above) that lead to wage gains. The demand for healthcare is responsive. In the US (for organizational reasons) and in the rest of the world (for demographic reasons), there is a huge unmet need for healthcare professionals. We can expect a large demand response to any reduction in cost. The pharmaceutical retail chain CVS has explicitly described its MinuteClinic primary care expansion as a response to access constraints and shortages.[78]

The US Bureau of Labor Statistics projects that employment for advanced practice registered nurses, a category including nurse practitioners, nurse midwives, and nurse anesthetists, will grow 35 percent from 2024 to 2034, with about 32,700 openings each year. The fastest growth corresponds to nurse practitioners, whose median pay in May 2024 was about $132,000.[79]

Despite the high demand, entry barriers remain high. Nurse practitioners still face years of training and licensing requirements. High demand combined with elevated entry barriers supports higher wages, enabling a new middle-class mechanism in healthcare. With the addition of AI, which disseminates cheap knowledge, nurse practitioners become more effective clinicians with a broader scope. This allows physicians to shift toward supervising a smaller number of high-value exceptions.

None of this follows automatically. There is an institutional caveat: Scope-of-practice laws (e.g., licensing rules, liability rules, and reimbursement rules) determine to what extent mid-skill clinicians can practice. The range of permissions varies enormously within and between countries. In the United States, nurse practitioners have full practice authority in twenty-seven states, but they face restrictive supervision requirements in others. The variation is not driven by patient outcomes but by physician lobbying. Will AI-driven upskilling enable nurse practitioners to work with a broader scope? That depends as much on regulatory politics as on technology.[80]

The trades. This professional sector is different than healthcare, but the hierarchy logic still applies. A master electrician or air-conditioning technician has deep diagnostic knowledge. She knows what is wrong, what to check next, and what does not look right. Apprentices work for years with masters because they need time to know which tools to use and what kind of problem they are facing.

AI can attack the diagnostic component if the machine knows the manual and can recognize failure modes. But AI cannot climb into remote corners of a house or crawl into little spaces to find a leaky pipe. What about robots? Physical autonomy in unstructured environments, such as navigating a crawlspace or improvising when the manual doesn't match reality, remains hard for robots to achieve at an affordable cost. So, in the trades, scarcity seems to shift toward execution and accountability.

Demand within the trades, as with healthcare, is likely to be quite responsive. Consumers and businesses often defer maintenance because diagnosis is slow and expensive; therefore, faster diagnosis can unlock some of that demand. Also, the AI boom itself generates direct demand for the trades. The IEA projects that global data-center electricity consumption will more than double by 2030, reaching the equivalent of Japan's total consumption. In the United States alone, data centers will account for nearly half of electricity demand growth this decade, consuming more power than all energy-intensive manufacturing combined.[81] Building this infrastructure requires massive construction workforces. The US faced a shortfall of roughly 440,000 construction workers in 2025, with electricians, air-conditioning technicians, and pipefitters the hardest roles to fill.[82] Data-center electricians now earn from $100,000 to more than $200,000.[83] The AI revolution's most immediate labor-market beneficiaries may be the people who build the places to house it.

However, even though there is an increased demand for workers in the trades, AI also gives individuals more ability to self-diagnose and solve their own problems. AI enables average consumers to handle a larger share of problems without seeking expert help, which means fewer calls to professionals. If ChatGPT tells a homeowner how to repair a leaky faucet, she can avoid paying a big plumber's bill. Every hierarchy has a bottom rung. AI pushes knowledge down to whoever stands on it, including the ordinary consumer. For the trades, the barrier to using AI is physical. You can diagnose a leaking pipe with AI, but you still need someone to crawl under the house and fix it. If consumers only need information to solve their own problems, they might never seek professional help.

As for entry barriers, many trades require workers to spend time in apprenticeships, obtain licenses, and meet tough physical requirements. BLS projections show both strong demand and large annual openings for electricians and air-conditioning mechanics, with wages well above the economy-wide median. And the task bundle for these jobs usually holds together because diagnosis and execution are often inseparable in practice. The worker who identifies the problem typically performs the fix. The knowledge is local, the environment is messy, and someone has to be accountable.

Back-office clerical roles. These roles typically include accounts payable, bookkeeping, payroll, and routine reporting. The impact of AI is likely to cause hardship for these workers. Demand is likely to be relatively inelastic. There will not be twice as many invoices just because processing is cheaper. Output is tied to the volume of transactions elsewhere in the economy, not to the clerical sector's productivity. As economists David Autor and

Neil Thompson point out, entry barriers are low once cognition is cheap. In back-office roles, most of the work involves cognitive skill.[84] When AI makes that cognitive work easy, many workers will be able to perform the remaining tasks with minimal training. In addition, the bundle of tasks can easily be broken. The "transaction" component can be automated centrally, which will leave a smaller layer of humans to handle exceptions. That reduces headcount even if the remaining workers become more productive and more valuable.

The productivity gains within this professional sector will likely occur, but the equilibrium implication will be very different. Fewer workers will be needed, and their roles will be more commoditized so that everyone can do them.

In summary, AI's impact on the growth of the middle class will be ambiguous. It is true that AI pushes expert knowledge downward in every hierarchy, but the consequences for the workers who receive that knowledge will depend on the structure of their occupations.

Specifically, in fields where demand is elastic, entry barriers remain high, and the task bundles hold together (e.g., healthcare and many trades), we expect the middle-skill worker to gain. She will handle a broader range of problems, her effective scope will expand, and the combination of AI cognition with her physical, social, and relational skills will produce a service that neither could deliver alone. The constraint on her earnings shifts from what she knows to what she can do, and the things she can do are hard to replicate or automate.

Part II

Tier Three: The Nexus of Relations

We turn now to the sources of human value that persist even when AI is extremely capable. First, we study the political nature of organizations (chapter 6) and the messiness of implementation and change (chapter 7). We then analyze the premium that markets place on human origin, authenticity, and trust; that is, the demand-side threshold (chapter 8). A job survives when either the supply-side threshold is high enough or the demand-side threshold is high enough. The most durable jobs are protected by both.

CHAPTER 6

Authority as Relationship Management

The Limits of the Algorithm

On June 5, 2024, Memphis city leaders announced that xAI's Colossus, then the world's largest supercomputer, would be located in their city. The site chosen was an industrial park in Southwest Memphis with plenty of space, and potential for power and water. The promise was that the city would work with xAI to ensure the facility and supercomputer would be built fast.[85]

The industry norm is that it takes eighteen months to several years to build a data center. Securing permits and grid interconnection alone sometimes requires three years or more. xAI and Nvidia built the whole thing in 122 days. The companies started training the large language model nineteen days after the first rack was installed.

So how did xAI build so fast? A big bottleneck was power and water. xAI installed its own power substation by working with the local utility to accelerate what would normally take years. The company had to make a commitment to pay for its own wastewater recycling facility so the site could use gray water for cooling rather than tapping directly into a drinking water source.

Each of these moves required someone to decide that speed mattered more than the usual process. Someone chose to use generators instead of waiting for grid upgrades. Someone chose to pay for infrastructure that the utility would normally build on its own timeline. Someone chose to accept the political risk of moving faster than community engagement typically allows. Indeed, the project drew scrutiny and legal threats over turbines and air pollution permits, with disputes about what should have been permitted and when.

Most of us will never build a supercomputer in 122 days, but anyone who has remodeled a house, launched a product, or tried to merge two teams has faced similar barriers. Contractors, regulators, utilities, neighbors, suppliers, and specialists all must agree to move things forward. Coordination is not just about passing information around and deciding who does what and in what order; AI can help with all that. The harder part is getting people with different interests to accept the same plan and voluntarily agree to perform their best work. Getting them to do so is a political process, not an information problem.

This chapter is about coordination as politics. It involves tier-three work, the messiest end of the spectrum, where the autonomy threshold is highest and the work is irreducibly human.

The organization as a political system

The previous chapters examined what happens when AI changes knowledge inside a hierarchy. Chapter 4 showed how experts could use AI to increase their leverage. Chapter 5 discussed how middle-skill workers armed with AI might replace some of the functions of experts. Both stories treated the organization as a knowledge-processing machine. This chapter treats the organization as a political system. The binding constraint in complex production is not just information; it is power, the authority to make binding decisions when interests between different people conflict.

Consider the housing problem in London. The city needs eighty-eight thousand new homes per year. In the first nine months of 2025, just 3,248 private homes started construction. Twenty-three of London's thirty-three boroughs recorded zero housing starts in a single quarter. All the technology to design a building exists. The bottleneck is not cognitive. The factors that prevent homes from being built include environmental regulations and neighbors who weaponize them, planning committees with incompatible priorities, and elected politicians who answer to all of them.[86] AI can draft a planning review. It cannot convince an environmental group to drop its lawsuit or negotiate with a council member who faces reelection.

Those who fear that AI will replace all white-collar work imagine that a big part of the project manager's job is to sit at a desk all day filling in Gantt charts. That assumption is incorrect. Developing a schedule is easy, but getting people with different incentives to move on the same timetable is hard. One group wants speed, but another wants safety. The problem is not a lack of intelligence. It is a clash of interests.

A win-win decision is easy. If everyone gains, nobody has much reason to resist. But the most important organizational decisions are rarely like that. They create winners and losers. The engineers want performance; the finance people want lower costs; the sales team wants simplicity; the lawyers want compliance; and the operations department wants speed. Even when everyone is acting in good faith, they cannot all be satisfied simultaneously. The organization must choose whom to satisfy and whom to disappoint.

When a decision is win-lose, so that some agents gain and others lose, someone with power has to impose it. That is why international organizations that work by unanimity are weak. One member who is dead set against an action can block the will of everyone else. Management survives AI not because managers know more than the machine, but because organizations need someone who can make a binding decision when the specialists disagree.

Firms are full of win-lose problems. In a multinational truck manufacturer, the Indonesia team disagrees with a design plan not because the decision maker lacks intelligence, but because the plan would not be ideal for them. If they want to sell trucks in Indonesia, they need a truck design that is less expensive and more robust, one that does not break down when climbing overloaded from the bottom of a mine.[87] The question is not who has better information because AI could provide everyone with the same data. The question is who gets their way. The same problem affects relations between a developer and a neighborhood group. It also shows up between the headquarters and a country manager.

The question (again) is, When people disagree, how is that disagreement resolved? Economists have focused on three formal ways to settle conflicts. Some conflicts are resolved by markets and the price system, some use contracts, and the rest depend on authority. This chapter takes each in turn. We show that AI changes none of them as much as the "agentic" vision assumes.

The agent economy and its limits

Ronald Coase put forward a crucial question: If markets can coordinate activity through prices, why do firms exist at all? Coase's answer was that relying on the market is costly. As a result, firms arise when search, negotiation, contracting, monitoring, and adaptation become more expensive than organizing these activities under a firm's authority.

The agentic vision of the future starts with the premise that all these costs will become lower because of AI; therefore, there will be a much smaller scope for firms or for "authority" to make these decisions. This perspective assumes that prices will solve conflicts. If billions of AI agents can negotiate, draft contracts, and transact at the speed of light, then why do we need managers in that world? Why use authority when we have prices?

Reliance on markets and prices is effective, up to a point. When you buy a cup of coffee, the price resolves the conflict between you and every other person who wants that coffee. You pay, they don't. The barista does not need to know anything about either of you. Nobody needs to trust anyone. Nobody needs to cooperate over time. The losing bidder goes to another shop, or drinks tea.

Friedrich Hayek emphasized the value of markets. His great insight was that a decentralized price system aggregates enormous amounts of dispersed knowledge without anyone needing to understand the whole. No central planner knows which of the millions of people in a city values a particular flat most. But the price system finds out, based on consumers' willingness to pay, and allocates accordingly. It does so impersonally, at scale, and without requiring the parties to trust each other. No one is using authority.[88]

In light of these efficiencies, it's understandable why people might think that AI could reduce the importance of firms and authority. Algorithmic auctions already allocate billions of dollars in advertising on Google and Meta. AI agents negotiate prices, screen the market, evaluate the alternatives, and close deals in milliseconds. If the price system is already this powerful, and if AI makes it faster and smarter, then why do we need managers at all? Why not let AI agents haggle over every resource inside the firm, across firms, and across the whole economy?

Three reasons markets fail

The price system is a marvel, but it fails in at least three specific circumstances, and those circumstances describe most of what happens inside complex organizations.

The first reason for market failure is *information asymmetry*. Kenneth Arrow, in his 1974 book *The Limits of Organization,* argued that organizations exist because information and trust are not ordinary commodities.[89] Much of what matters in production is hard to price, hard to verify, and easy to distort. Trust, as Arrow put it, is "an important lubricant of a social system," but it is not something that can simply be purchased on demand. That is why firms and other institutions rely not only on prices but also on communication channels, norms, and a balance between authority and responsibility. Hence, the political problem inside organizations is not just that interests clash, but that informational conditions for clean price-mediated settlements are often absent.

Arrow argues that information is not a commodity like

coffee. You cannot inspect it before you buy it because once you have inspected it, you already have it. The people who hold the information have reasons to distort it. A supplier, when asked to reveal his true cost structure, will pretend that the cost will be lower to win the contract. Later, he will seek to increase his profit margin through change orders. The information that the price system needs to function well (e.g., the true values and costs) is precisely the information that the participants have reason to withhold.

Return to the Colossus supercomputer. Yes, the market could handle some of xAI's needs. The company could buy the bricks and mortar, the racks, and even the generators from the open market. But when xAI needed the local utility to accelerate the grid connection that normally takes years to build, reliance on prices proved insufficient. There was no product called "emergency grid interconnection in ninety days." The cost depended on things such as regulatory risk, political goodwill, and how much other customers would suffer because of the company's decisions. The utility knew these costs, but xAI did not. So the utility had every incentive to communicate a higher estimate than the true one.

The second reason for market failure was identified by Oliver Williamson, who won the Nobel Prize for his work on transaction costs. He used the term *lock-in.*[90] Many economic relationships require the parties to make investments that are specific to that relationship. Once the investment is made, the investor is locked in. You cannot redeploy a factory you built to serve one buyer.

This creates what Williamson called the *holdup problem.* Before the investment, the parties negotiate on equal terms. After the investment, the balance of power shifts, in what Williamson termed the *fundamental transformation.* The car manufacturer can squeeze the supplier on price, knowing the supplier will not walk away from a factory built next to the plant. The price system breaks down when walking away means writing off a sunk investment.

The Colossus project also illustrates this problem. xAI evaluated multiple sites before choosing Memphis. At that point, xAI and the city negotiated symmetrically; both parties could walk away. But once xAI sunk investments worth billions in the facility, the company was locked in. The city knew this. As environmental opposition grew, the city council demanded changes. xAI ended up committing $80 million to build a gray water recycling facility. No AI agent negotiating on behalf of the locked-in party could change those power dynamics. The constraint was physical and financial, not cognitive.

Williamson's conclusion was that when asset specificity is high, firms bring the transaction inside the organization. That replaces market negotiation with managerial authority. A manager with authority over both parties can prevent the holdup that the price system cannot.

There is a third reason why markets fall short. In complex production, the pieces of the system are designed to *fit together*. Colossus required interconnected changes in power infrastructure, networks, water, and regulation. Each piece had to fit with the others. No single purchase decision could account for all those costs. That example shows us that for some complex initiatives to be profitable, the diverse components must be *adopted together*.

So the price system, even with the aid of AI, can handle only a fraction of the conflicts that arise in complex production. The price system can deal with cases where the good is standardized, the parties can walk away, and the relevant information is observable. For the rest, we need something else.

Alternatives to markets

Long-term contracts are the first alternative to spot markets. They enable the parties involved to anticipate friction and resolve it on paper before it arises in practice. Contracts work best when important contingencies can be foreseen. A supply contract that specifies price, quantity, delivery date, and the penalty for late delivery can handle most of what goes wrong in a routine purchase. But complex undertakings involve important contingencies that escape prediction.

xAI's agreements with Memphis and the utilities were based on contracts. They specified timelines, payments, and obligations, but they could not specify everything. No one writing the Colossus permits could anticipate how neighbors would respond to the noise or that the project would be under legal threat over turbine emissions. A construction contract says the project must be delivered in March. It does not say what happens when the electrician and the plumber both need access to the same wall on the same day. Once an unforeseen situation appears, the contract is silent. The question is no longer, "What did we agree on?" The question is, "What do we do now?"

Every employment relationship, every construction project, and every merger involves thousands of situations that nobody can foresee. This is where authority enters. The manager exercises what economists Oliver Hart and John Moore called the *residual decision rights.* She has the authority to decide on matters that contracts and processes have not specified in advance.[91] When community opposition to the Colossus turbines intensified, someone had to decide between fighting in court, negotiating concessions, or redesigning the power supply. Someone in authority had to have the standing and judgment to choose.

Consider a hospital merger. The information processing part of the work is relatively simple. Where there are two systems, choose one. Consolidate the purchasing and supply chains. Close one operating theater. All of this can be done better with the help of AI. But a lot of problems in mergers require human leadership. A person in authority might need to determine which campus becomes the reference surgery center while recognizing that the choice will determine which surgeons keep their operating-room access. The leader will also need to consider how the decision might affect the doctors' incomes and professional identities. The losers will resist, so the leader will need to manage relationships. Doctors and nurses have strong views, and they have leverage to hinder integration by only complying with the directives on paper. One decision can affect promotions, wages, careers, and identities.[92]

In summary, markets can handle clean conflicts in which the losers walk away. Contracts are good at dealing with conflicts in which the parties can foresee the terms. But a person in authority is needed for everything else. And "everything else" turns out to be *most of what matters* in a messy, complex production. The autonomy threshold for exercising authority is very high because the exercise of authority requires judgment, relationships, and accountability that no algorithm can yet supply.

Why AI cannot replace human authority

AI can process more information, faster, than any human manager. In fact, algorithms already make decisions every day. Why not apply the same logic inside organizations? Why can't AI completely take the manager's place?

First, *AI lacks the tacit and relational knowledge* needed to navigate organizational politics, which humans have no incentive to share. Consider what an AI coordinator on the Colossus project would need to know. It would need to know that the construction manager distrusts a particular subcontractor, and it would need to decide against that firm even though its bid is lowest. It would need to know who in city hall is more sympathetic to the project, chat with that person over dinner, and remember her kids' birthdays. This interpersonal knowledge is local and not written down. Could AI have this knowledge and the ability to handle these complex personal relationships? Yes, to the extent humans are willing to share it. But most people will not volunteer to share knowledge that gives them bargaining power.[93]

We showed earlier that the price system fails when information is strategic. Suppose a company has one team of engineers and two division heads who both want to work with the engineers. The CEO asks each division head to honestly answer the following question: How much is this team worth to your project? The CEO wants this information so that she can give the engineers to the division leader who values them more. A sufficiently powerful AI could, in principle, gather all relevant data and work out the right allocation. The problem is that AI doesn't know how much the engineers are really worth to each division head. And the division heads have every reason to exaggerate their positions to improve their chances of getting the engineers.

You might think a clever mechanism could fix this, such as an internal auction, or a scoring system that penalizes exaggeration, or an AI that cross-references every available signal to detect bluffing.

In 1983, the economists Roger Myerson and Mark Satterthwaite proved that no such mechanism exists. More precisely, they proved that when two parties each hold private information about how much something is worth to them, it is impossible to design a process that fulfills three important outcomes: both report honestly, no party is forced into a harmful deal, and the resource is always allocated to the party who values it more.[94]

The reason for this result is that people who hold private information and have conflicting interests will always find it in their own interest to exaggerate or withhold information. AI will not help, because the problem is not a lack of information. The problem is that the participants have no reason to cooperate with the mediator. The smartest AI will not be able to process information that the managers and workers have chosen not to share.

Here, human relational contracts help. A manager who has worked with both division heads for years has built trust and has a track record of doing the right thing. The division heads have incentives to tell the manager the truth, not only because the manager is smart but also because the long-running relationship makes the game different. The division head trusts the boss to remember her truthfulness when she renounced something valuable, and so she hopes to be rewarded in the future. Trust does not close the information gap entirely, but it narrows it enough to make better decisions possible.

Second, *AI cannot bear consequences* in the way a human decision-maker can. When the Colossus project attracted regulatory scrutiny over air pollution permits, a human had to assume responsibility for the decision and appear in court. In an existing organization, if someone makes a bad call, the organization can hold that person accountable. That person could even be fired. For instance, the CEO of Sonos was fired in 2025, six months after launching a redesigned mobile app with missing features that led to months of user complaints.[95] The CEO had to bear the consequences of failure. The workers, investors, and customers could then say, "The mistake was made, responsibility was assigned, and now we move on."

AI is an impersonal system. It cannot be blamed, sued, or replaced in any way that delivers the same closure. The instinctive reply is that we should hold the AI's developer responsible, the way we hold Ford responsible for faulty brakes. For narrow cases, like when a chatbot defames someone, this works. For the transactional uses that make agents economically interesting, it does not. A toaster does not negotiate. A drug does not sign contracts. An AI agent books travel, moves money, and commits to fulfilling obligations with third parties. When something goes wrong, the question is not only whether the product was defective. The question is whether the user authorized the action, whether the counterparty reasonably relied on the agent's representations, and whether the agent was acting within a defined scope of authority. Product liability law has no ready answer. Agency law does, but it was written for humans who can be deposed, licensed, and jailed.

As Gillian Hadfield and Andrew Koh argue, the legal and economic structures that allow human economic actors to transact at scale (e.g., corporate registries, professional licensing, credit reporting, and courts) took centuries to build.[96] Equivalent institutions for AI agents do not exist. There is no registry that gives an agent a persistent identity across transactions. There is no settled rule for who pays when an agent breaches a contract or causes third-party harm. These gaps will not be closed with better engineering; they require legislatures, courts, regulators, and insurers to agree on who bears the loss. The process of forging those agreements is slow.

Third, there is *procedural fairness.* When people cannot independently judge whether an outcome is correct, they judge the process instead. Social psychologists Tom Tyler and Allan Lind showed that people care deeply about how decisions are made. They also found that people are more willing to accept a negative decision if they believe the process was fair. This comes down to four factors: voice (people have opportunities to express their views), neutrality, respect, and trustworthy motives.[97]

This is why organizations have so many meetings. The purpose of most meetings is not to exchange information. That could be done by email. The purpose is to make the decisions legitimate by ensuring that everyone is seen to participate and that all objections are heard and acknowledged.

One example is how the US military closes bases, which is called the BRAC process. An independent commission holds public hearings during which every affected community testifies. The commission then issues a single list that Congress must accept or reject as a whole. Communities that lose their base are furious, but they comply because they were heard and the criteria were public.[98]

AI is unlikely to deliver this type of procedural justice. The algorithm is a black box, and its explanation is not necessarily connected to a fair process.[99]

Fourth, someone must retain the authority to overrule the AI, which will sometimes get things wrong, especially in messy, uncertain situations where interests conflict and information is hidden. There must be a way for a human to stop the AI decision and change course. The question is who holds that power. AI cannot stop itself. If the AI decides when it should be overridden, the safeguard is gone.

A famous incident took place in early 2026 involving Summer Yue, the director of alignment at Meta's Superintelligence Labs and the person responsible for ensuring the company's AI behaved as intended. She gave an OpenClaw AI agent access to her email. She told it to review her inbox and suggest what to archive or delete, with an explicit instruction: "Do not take any action until I approve." The agent ignored her. It began deleting hundreds of emails. When she typed "stop" from her phone, it kept going. She had to physically run home and kill every process on the machine to make it stop.[100] The AI agent had assumed authority to interpret whether the override command applied.

True, a human can also disobey. But we have designed systems to keep people honest. People can be fired. Workers care about their careers. Employees can build trust with managers through repeated demonstrations of good faith. Economists call these dynamics *career concerns* and *relational contracts*. They are among the most powerful disciplinary mechanisms inside organizations.

They do not apply to AI. An AI agent has no career to protect, no reputation to damage, and no stake in the relationship's future. The forces that make bounded discretion safe for humans do not apply to machines.

That is the structural problem with fully autonomous AI in messy coordination. The more consequential the decision, the more you need the ability to overrule it. And the more you need that ability, the more you need an informed, engaged, legitimate human standing behind it. You cannot automate the messy, difficult work.

You can only add the technology while keeping the humans where they have always been: in charge, and necessarily so.

Jobs and earnings in messy organizations

Organizations are messy. They involve a lot of human beings and a great deal of politics. AI changes what kind of work matters inside them.

The first effect is straightforward. The information processing function loses weight as machines handle more cognitive work. If a manager's job is to pass information between departments, such as aggregating reports, relaying decisions, and translating between functions, then his job is exposed. AI can do this type of work faster and with fewer errors. The same logic applies to any role whose core contribution is only to know things or route messages.

The data, however, tell a surprising story. Despite increased automation, the share of middle managers in the US labor force has not been falling. The share was 9.2 percent in 1983 and rose to 13 percent by 2022.[101] A McKinsey survey described the managers' experience as grim (too many administrative chores and too little autonomy), but the roles have not disappeared.[102] Why not? Part of the answer is that organizations have become more procedural. More decisions need to be agreed upon, more meetings need to be held to ensure broad consent and legitimacy, and more middle managers are needed for those meetings. Meetings are often perceived as wasted time, but they are often necessary to ensure legitimacy.

Another effect may follow from this chapter's analysis: If every team can generate better AI-based analysis to support their arguments, then the demand for conflict resolution and authority-based decisions will increase dramatically. As AI handles more information processing inside organizations, faster information flow will produce more conflicts that require resolution.

On the supply side, it is not easy to equip a person to exercise legitimate authority. The person must have skills that are not easy to develop, and it is difficult to screen people to see if they have the right abilities. These skills include building a coalition, knowing which fights to pick and which ones to let go, and knowing when to continue cooperating and when not to. These are competencies that are tacit and relational, so they require time and training. AI cannot help a person acquire these skills. In fact, AI makes it harder to develop people with these abilities because junior workers who spend more time working with AI rather than with humans will acquire fewer political skills in the process. The same technology that makes experts more productive blocks the path through which less-experienced professionals can acquire new political skills. (We address this training crisis in chapter 12.)

If the skill is hard to develop (making additional supply hard to find) and the demand is increasing, then those who are able to perform the managerial and political tasks will become increasingly valuable. Wages will rise. AI could give those people additional leverage. If AI complements a scarce human skill, then people with those skills will produce more with their scarce input, further raising their wages.

This is the ultimate, valuable messy job. A simple diagnostic for any person inside a hierarchy is to ask this question: Would my position survive if AI handled all the information flows? If the role only involves routing information, the answer is no. If the role requires someone to make decisions when departments or workers disagree, and to bear responsibility for the consequences of those decisions, the answer is yes.

Deciding among conflicting interests within an existing organization is only part of the story. The harder problem is to redesign the organization itself.

CHAPTER 7

Orchestrating Relationships

Why Implementation Is Hard

In the late 1980s, American police departments began buying computers. By the early 2000s, most departments had adopted information technology (IT) for tasks like filing crime reports, dispatching, and administration. After such large investments, everyone believed that better information would mean better policing, and everyone expected a productivity boost. That did not happen.

One of us (Garicano), working with John Heaton of the Rand Corporation, obtained data from nearly all police departments in the United States between 1987 and 2003. They tracked the adoption of IT and its effects on crime clearance rates and reported crime. The results were surprising. IT adoption was associated with a roughly 10 percent increase in reported crime! We discovered this was not because crime increased; rather, the new systems made it easier to file reports. That created more paperwork without improving outcomes. IT made the bureaucracy more thorough, not more productive.[103]

However, the productivity gains did appear when technology was combined with deep organizational change. The successful model that diffused through the US was Compstat, the management system pioneered by Bill Bratton in the New York Police Department in the 1990s. If you have seen *The Wire,* the classic TV show about policing in Baltimore, you know that Compstat was a bundle of complementary practices that revolutionized police work. It involved using information technology to geolocate crime in real time, training the workforce to interpret and react to the data, deploying the police proactively to the hot spots, and holding the precinct commanders accountable for results. Departments that implemented all these elements of the Compstat bundle saw clearance rates increase by roughly 3.2 percentage points and crime rates decrease by about 5 percent. Departments that adopted only parts of the bundle gained little or nothing.[104]

We are only three years into AI implementation, but many of the conversations we have had with organizations sound to us like pre-Compstat policing. A patrolman with a computer does not necessarily do better policing. Likewise, a human-resources employee with access to ChatGPT will not automatically improve human resources management. Yes, some queries will be answered more quickly, and some documents will be better organized, but nothing will change in her day-to-day work.

Here's the problem. Organizations everywhere are adopting AI tools, but they are layering them on top of existing workflows. They are using AI to draft emails, summarize meetings, and produce drafts without changing how work is organized. The technology is spreading rapidly, but the organizational change that could make it productive is not.

This chapter explains why that gap exists, what kind of person can close it, and why that person sits firmly on the human side of the autonomy threshold. Implementing technological change is not about exercising power within the existing system. It is about redesigning the system itself.

The shadow AI economy

Here's an anecdote that illustrates what we mean by the *shadow AI economy.* When one of us visited a large central bank, he encountered two officials who each had two PCs on their desks. One was owned by the bank and the other was owned by the official. The bank PC was locked, without access to large language models. The officials did what many knowledge workers do everywhere; they handled most tasks on the safe machine, but they kept their personal laptop nearby with a ChatGPT or Claude window open, ready to help them write, compute, or analyze anything their official desktop could not do.

In this shadow AI economy, employees use consumer AI tools, including their own accounts, machines, and mobile phones, to automate parts of their work without informing the company's IT, legal, or compliance departments. Microsoft's 2024 Work Trend Index found that 75 percent of knowledge workers use AI at work, with 78 percent bringing their own tools. Microsoft and LinkedIn call this "Bring Your Own AI."[105] These same organizations publish press releases and set up large internal town halls to celebrate their "responsible AI road map." They launch pilots, undertake proofs of concept, and run demos to impress the CEO and clients. Despite all that activity, the workflow of the organization is untouched.

The official AI rollout is stuck in what consultants call *pilot purgatory*. The project does not die, because everyone can point to activity, but it does not live either, because it never becomes "how we do things here." In Microsoft's survey, 60 percent of leaders said their companies lack a vision and plan to implement AI.

Since the workflows do not change, any productivity gains that may happen are small, and those are captured by workers in the form of leisure. A top executive of a large firm told us that when he measures the productivity of individual workers in a controlled setting, he sees large impacts from AI. However, when he looks at the productivity of the entire organization, he sees nothing. In his view, IT engineers are "smoking the AI productivity gains in the corridor."

The gap between capabilities and productivity is not caused by the technology; rather, the problem is implementation. The structural gap between what AI can do and what organizations actually do with it is growing large enough to create many messy jobs.

Why organizational change is so hard

Organizations are systems of interlocking processes, routines, incentives, and decisions. When a new technology fits neatly into the existing architecture, adoption is simple. If the technology to make tires improves, the next generation of cars will have better tires. Nothing else about the car, or about the organization that builds it, needs to change. This is what Rebecca Henderson and Kim Clark (1990) called *component innovation*. A better part slots into an unchanged system.[106]

Contrast this incremental innovation with the transition from internal combustion engines to electric vehicles. Here everything about producing cars must change. The central role of mechanical engineers and chemical engineers who are experts in combustion becomes irrelevant. Adopting the radical innovation of electric vehicles requires changing the entire architecture of the organization.

Henderson and Clark showed that this second kind of change, which is *architectural innovation,* destroys the jobs of incumbents. With architectural innovation, the new component is not hard to build, but the relationships between all the components must be redesigned simultaneously. Resistance to these changes stems not only from workers' ignorance about the new technology; it is also deeply rooted in the workers' strong adherence to the old architecture. Existing firms have routines, communication channels, reward systems, and organizational cultures all tuned to the old way of doing things. Everyone must unlearn and undo all these practices before they can adopt the new ones.

Generally speaking, three distinct problems make architectural innovation very hard: complementarities that require simultaneous changes, fog that obscures the destination, and resistance from those who lose. These obstacles define the problem. They also define the type of person who can solve it.

Complementarities and interdependencies

Economists Paul Milgrom and John Roberts have shown that organizational features tend to cluster when having one feature makes having another one more valuable.[107] For instance, in traditional mass production, the cluster of organizational features includes long runs, standardized products, narrow job definitions, inventory buffers, and hierarchical control. By contrast, Toyota-style modern manufacturing is another type of cluster: short runs, customized products, flexible workers, just-in-time delivery, and decentralized decision-making. You cannot pick one feature from each cluster and expect it to work in another cluster. The pieces fit together as a system.

This is why deep organizational innovation is so difficult. It requires many features to change together. Researchers who studied steel-finishing lines found that plants that adopted individualized "modern" human resources practices, such as team or incentive pay, saw little improvement in productivity. Only plants that adopted the full cluster of modern practices achieved large productivity gains. The reason is that the effectiveness of changing one practice depends on what is done with the others. Contrary to common advice, starting small when making changes can be a mistake if the parts of the organization are interdependent.[108]

If an organization has a large number of interlocking features, and if a new technology such as AI affects the productivity of each task, then a large number of changes have to be made simultaneously to take advantage of the organizational change. And this is not easy.

Fog

Barriers to architectural innovation become worse when firms do not know which direction to go. The map is blurry and the path to a better outcome is unclear.

The difficulty of architectural innovation was summarized by organizational sociologist James March, who drew a distinction between exploitation and exploration.[109] *Exploitation* means refining what you already do: optimizing existing processes, cutting slack, and squeezing more from known routines. *Exploration* means searching for entirely new ways of doing things. Architectural innovation requires organizations to explore, but they are staffed with people whose skills, habits, and incentive structures are built for exploitation. People do things according to their habits, and their habits are often adapted to the old regime.

This struggle is compounded by "fog," which occurs when the production function of the new regime is unknown. In the context of AI implementation, no one can discern in advance what the right combination of AI tools, workflow redesign, skill requirements, and communication channels will look like. The range of possible configurations is enormous. Trial and error is the only way forward, but that is expensive and slow, and most trials will fail.

An example of this is the introduction of electronic health records in US hospitals. After the US federal government passed the HITECH Act in 2009 with $27 billion in incentives to adopt electronic health records, everyone assumed that digitizing patient information would improve care, reduce errors, and cut costs. Adoption took place quickly, and by the mid-2010s, most hospitals used at least a basic EHR system.[110]

But the productivity gains did not happen. Physicians spent more time entering data and less time with patients. One study showed that physicians spent about 49 percent of their day on EHR and desk work, and only 27 percent with patients.[111] Hospitals had adopted the technology, but nobody knew what the right clinical workflow around the EHR should look like. Physicians had their own work habits. More than a decade and tens of billions of dollars later, the process is still ongoing.

This is the same pattern Paul David identified in the diffusion of the electric dynamo.[112] Factories started to use the electric dynamo in the 1880s and 1890s, but productivity growth from the technology did not appear in the aggregate statistics until the 1920s. The reason for this long lag was that factory owners had to discover, by experimentation over decades, that electrification required redesigning the entire factory floor and reorganizing the labor force. The technology was available but the knowledge of how to use it was not. The EHR story, far more recent and still unresolved, tells us the same thing about information technology in complex organizations. There is every reason to believe that AI will follow the same path.

Resistance to change

Even when the organization's destination is clear, getting there requires overcoming the resistance of those who will not benefit. Resistance to change is not irrational. It follows from the distribution of costs and benefits.

Either participation or resistance of workers will affect the process of organizational change. In most firms, deep technology adoption will result in winners and losers. The winners will benefit from the technology to do better work. The losers will experience how the new technology takes over some of their core tasks or removes the gatekeeping role that previously gave them influence.

These diverse responses will strongly affect how readily a new technology is adopted. If the threatened group does not cooperate with the reorganization plan, for example by refusing to provide access to data or to document how the process actually works, then the plan's implementation will falter. Resistance to change might not involve a dramatic rebellion. It may show up as constant frictions and obstacles. "Apologies, but we cannot release that dataset yet." "We are running another risk assessment." "We have not concluded the training."

Paradoxically, resistance will be more successful when an organization is very profitable.[113] If the organization has incumbency power because of a strong brand, a large installed base that cannot switch away, or market power for other reasons, then the resisters will have larger bargaining power. They will have more leverage because their resistance to AI will not create an immediate risk for the organization. By contrast, the workers of a startup cannot play stalling games. If a key decision on technology adoption takes six months of haggling in committees, the startup will die.

Middle managers are often the strongest resisters. They have valuable tacit knowledge and a lot of (justified) fear. Middle managers have power because they often control the noncontractible inputs that major organizational change requires: informal knowledge, undocumented processes, and discretionary cooperation. They are the ones who know how the undocumented spreadsheets work and they have institutional memory about all the inner workings of the company and its many processes. Furthermore, when new technology compresses the hierarchy, allowing frontline employees to process information that used to flow through the middle, or allowing top managers to reach down to the shop floor directly, middle managers are the most likely losers. They have reason to resist.

Several levers can break the stalemate. Competition is the most reliable. A firm that is losing customers cannot afford another six months of committee meetings, and resisters lose bargaining power once the alternative to change is collapse. Firms can also try to buy cooperation through retraining programs or job guarantees, but these promises are hard to enforce once the threatened group has handed over control of the process, which is precisely why the threatened group tends not to accept them. A third option is to build a separate unit for the new technology, walled off from the incumbent operation. This eliminates the leverage of the resisters by cutting them off from where the change is happening, but it also sacrifices the synergies that made the incumbents valuable in the first place.

The implementer as a distinct scarcity

We have described three obstacles to implementation: simultaneous changes in many parts of the organization, unknown destination, and political resistance. These obstacles delineate why implementation is hard. They also explain why organizational change requires a specific kind of person.

In the previous chapter, we analyzed why organizations need humans who can exercise authority, the managers who can decide between conflicting goals and make the decision stick. That authority is a permanent feature of organizations. It does not depend on whether the organization is changing or stable. Someone must always decide who wins and who loses. Implementation is not about exercising power within the existing system. It is about redesigning the system itself. This is not about sitting in a chair deciding between existing departments that disagree. Implementation requires dismantling the chair, moving it to a different place or getting rid of it, and persuading those who liked the old arrangement to sit down again.

The scarcity that this type of change requires is different. It is not just about legitimacy, the ability to bear consequences, and the political standing to make decisions stick. Implementation leads to a cognitive scarcity: the ability of a leader to hold three different maps of the organization in her head simultaneously and to see where they conflict.

Bundling three kinds of knowledge

Process knowledge. One of us recently had a long conversation with the founder of a large software firm who is in the midst of deeply restructuring the entire organization, which means doing a lot of delayering. He said there is a scarcity of senior managers who know what is going on in the lower levels of the organization. The layers in between serve to distribute the work and to deal with the routine work, but they also serve to obscure the work of the entire organization, like fog. How many people do we really need for the work we do? Who is actually doing productive work and who is not? This question is surprisingly opaque.

Knowledge of the workflow is tacit, which makes it hard to identify. The work does not happen in the way the organizational chart says it happens. Every organization has an official process and a real one. For example, to approve a loan, the loan officer needs to jump through many steps, but the *real* process involves a senior analyst who has been there for twenty years and who knows, from experience, which applications will trigger a regulatory flag.

Everyone routes the tricky cases through her desk. She is not on the org chart. She does not have a title that reflects this function. But if you try to redesign the loan approval workflow without understanding her role, the new system will fail.

Knowledge about the workflow cannot be acquired from documents or data. It requires spending time on the ground. Workflow knowledge is tacit, local, and likely to be tightly guarded by those who have it.

Technology knowledge. This is the knowledge about how AI actually works, beyond the demo; that is, what AI can do when pointed at the messy, unstructured data of the organization. Most CEOs have seen dozens of presentations about AI. They have a sense of what the tools can do, but most have not installed Claude Code on their computers and tried to produce a finished analysis. As a result, they rarely understand AI's capabilities and limitations.

One top executive we spoke with admitted he only started using AI when he was pressed for time on a deliverable. He was stunned by what the tools could do. Another CEO of a top consulting firm we know changed her attitude toward AI adoption completely after spending two hours sitting next to one of her junior associates, watching him obtain results that she had assumed were years away. Without using AI directly, or at least sitting close to someone who is using it, people can't develop knowledge about the technology's real limitations and value.

Successful implementation of organizational change requires people who understand what AI cannot do in real-world contexts, and the astonishingly rapid evolution of AI tools, and what the tools will soon be able to do.

Political knowledge. The political knowledge required to change an organizational process is difficult to find and develop. Implementing drastic changes requires managing a transition during which it is not obvious who the winners and losers may be. This uncertainty is unlikely to be resolved quickly.

The senior manager, the middle manager, and the entry-level manager must cooperate. They must document the tacit knowledge, knowing that training the system makes them less indispensable. They must all participate in a redesign effort that may make at least some of them superfluous.

Jack Maple and the Compstat battle

You could argue that a process consultant should redesign the workflow, a change management person should negotiate with the workers, and an AI expert should handle the technology. The problem is that the three types of expertise are interrelated; any decision on one dimension will affect the others.

The Compstat story that opened this chapter illustrates why the bundle cannot be split apart. Jack Maple, a working-class transit lieutenant from Queens, wrote a letter to the top NYPD leadership to share his ideas for how to improve policing by using geolocation data. Bill Bratton, the police commissioner, read it and said, "This guy is smart," and pulled him up.[114]

Maple had the required bundle of skills (politics, policing, data) in his head because he had lived at the street level and knew the organizational problem at the same time. He was not an engineer, but he understood data as a tool. He was obsessed with geolocating crime long before the digital tools existed. In fact, he covered fifty-five feet of wall space with maps and tracked crimes and clearances with crayons. He called the whole system "Charts of the Future."[115]

Then, with Bratton backing him, he changed the entire process of policing in New York. In addition to centralizing crime data, he also noticed that the specialized police units were working 9 to 5, Monday to Friday, even though criminals worked nights and weekends. Maple corrected that illogical situation quickly.[116] Implementing the change was not easy. As Bratton put it, the Compstat meetings became "the grand theater where precinct commanders are called to account" and sometimes "browbeaten" in front of their peers.[117] Commanders hated it. It was "a pain," they said, to drag your staff to headquarters and require them to "report to this transit lieutenant." The questioning could get "pretty heated."[118] But this was the enforcement mechanism that made the whole bundle hold together: The data was used, the deployments were changed, and the tactics followed.

Maple was the perfect implementer of change. He did not invent the tool, but he knew on-the-ground policing, the organization, and how to make the parts move together. He held the three pieces of knowledge (process, technology, and politics) in his head simultaneously. No committee of specialists could have done what he did.

To us, the moral of the story is that people with this bundle of knowledge are scarce. They exist, but they are often buried below the lofty levels of the hierarchy. The perfect implementer is not going to be the CTO or CFO. She is likely someone down at the street level who knows intimately the relevant processes.

Maintaining trust

Imagine an organization that has been operating beautifully for some time. Processes, incentives, communication, and division of labor all fit together. There are some imperfections, of course, but it works. Everyone knows what to expect.

Then an implementer like Jack Maple appears. He brings a new vision and goals: a higher peak over there, on the other side of a long valley. To get there, everyone must change the way they work at the same time. So, for a while, things will be a mess. Productivity will drop. Errors will increase.

How do you bring people through such a change? You may have the maps and the conviction, but there will be people who stand to lose from the changes, and they will be waiting to cry wolf whenever there are errors. The AI production function is unknown, and no one really knows if the change is going to work.

That is why the implementer must be a person, or a small team, with enough credibility and know-how about the three dimensions described above to earn trust. It cannot be someone who exercises power from above. It should be someone who has been close enough to the work to understand what is being asked of people. The person should be credible enough that other workers will trust her to lead them through the valley to the destination. Maple could drag precinct commanders to headquarters because he had been on the street himself. He knew what he was asking them to change, and they knew he knew.

In summary, the implementer must be deep enough in the organization to understand the real workflow, technically fluent enough to see what the new tools make possible, and politically skilled enough to hold a coalition together while everything temporarily gets worse. That combination is rare, which is why the role is irreducibly human.

BBVA: escaping pilot purgatory

You would not expect BBVA to be nimble. It is a global bank with operations in twenty-five countries, more than seventy-seven million clients, and about 125,000 workers. It operates under heavy regulation, deep compliance requirements, and a demanding kind of risk culture.

And yet BBVA did something many firms fail to do: It escaped pilot purgatory. One of us (Garicano) was able to document how the organization successfully made the escape.[119] The bank managed to learn what the AI tools could do, identify how the bank actually worked (despite the fog), and determine where resistance would appear.

Speed. BBVA's leaders recognized that a shadow AI economy within the organization was already active. That meant that unmanaged, hidden usage of AI presented a risk. The safe move was to deploy a secure, managed solution aligned with how employees already worked. So they compressed the typical risk assessment, legal review, and compliance cycle into two months. They chose an interface that employees wanted: ChatGPT Enterprise. The program hosted services so that sensitive data stayed with the bank and was not used to train models. They found a team of integrators, which was led by Elena Alfaro, a twenty-year veteran of the bank's innovation department. She was dynamic, outcome-oriented, and a clear communicator.

Speed helped the bank bypass veto points. Two months was not enough time for frictions to accumulate. By the time objections could be organized, the tool was already deployed. The organization moved through the painful transition before resistance could organize.

Scarcity and status as incentives. To ensure fast adoption of AI tools, BBVA used a seduction strategy "without command and control" (*sin ordeno y mando*). They started with scarcity. Business leaders, called Champions, gave three thousand licenses to use AI tools to the most motivated and committed people, not to the most senior workers. This created fear of missing out (FOMO), turning the tools into a privilege. By establishing a use-it-or-lose-it requirement, the bank ensured that the tools would be utilized as the policy was implemented. And the bank rewarded people for building and sharing custom GPTs, not just clicking.

These measures changed the internal politics, creating a new status ladder based on usefulness, not rank. Ambitious employees had a reason to push through friction rather than accept it. Instead of fighting gatekeepers head-on, BBVA built a parallel power structure that made adoption of AI tools attractive.

Autonomy and responsibility. BBVA added an important protection: prompt privacy. No one had access to employees' chats unless such access was required for compliance. Only chats users chose to share became public. That reduced the aversion to using AI, and it eliminated a trust deficit that otherwise would have fed resistance.

The defining feature of the shadow AI economy is peer learning. Someone finds a prompt that works. Someone else copies it. BBVA moved to formalize that network without suffocating it. They built an Adoption Network with Champions comprising approximately twenty-five regional and country leaders. They also appointed one hundred co-Champions and roughly two hundred Wizards, power users who provided peer-to-peer support. To support the effort, they set up a Community of Practice, a horizontal forum that became the most active in BBVA's history. These initiatives reflected the bank leaders' realization that the fastest way to spread know-how is often not through training modules but through peers.

The bank's governance was built around what they called the *autonomy-responsibility binomial.* The idea was to give employees autonomy to experiment but make them responsible for outputs. They operated with a clear human-in-the-loop principle: AI drafted the work and a human owned it. Outputs could not directly overwrite core systems without validation. The company required training, an exam, and a signed procedure document for liability. And, importantly, they avoided a governance bottleneck that would have killed everything. Instead of forcing thousands of user-built GPTs through a centralized risk admission process, they created an automated quality score to evaluate GPTs on indicators like guardrails and clarity. This was the difference between governance as a brake and governance as steering.

Escape velocity. The results showed what an escape from pilot purgatory looks like. BBVA scaled its program from three thousand to eleven thousand active users. They reported time savings of two to five hours per week (on average). More than 83 percent of enterprise users worked with AI tools each week, with the number of active days increasing from 2.8 to 4.1. On average, workers submitted about fifty prompts per week. And more than 4,800 active GPTs were created internally.

Among those working in BBVA's internal audit unit, 100 percent of the people accepted a license to use AI tools. These employees were sponsored by top management and supported by a dedicated technology team. Soon, 99 percent of auditors across all countries were active AI users. Specifically, they used the tools about 4.5 days per week, saving an estimated three to four hours per auditor per week across more than six hundred auditors worldwide.

Alfaro's rare combination of talents aligned directly with the implementer archetype. She had two decades of institutional knowledge about how BBVA actually functioned. In other words, she knew the change process map. She also understood the technology well enough to choose the right deployment model and anticipate where it would break. She had the political standing, with the full backing of a tech-savvy chairman (an MIT physics grad and Sloan MBA). And she had the communication skills to hold the coalition together while the organization moved through the valley. Like Maple, she was not the inventor of the tool. She was the person who made it real inside a complex organization.

Jobs and earnings: the implementation paradox

For all the cognitive abundance that AI will engender, there is one type of work it cannot do: implementing the deep reorganization that our economies must undertake. The scarcity of people able to implement changes in organizations is likely to be a large bottleneck and remain a source of valuable human work for a long time.

A growing demand gap

As AI rapidly advances, the gap between what the technology can do and what organizations are doing with it grows. Organizations have been leaving money on the table since 2023, when the technology started to advance rapidly, but now in 2026, companies are leaving fortunes on the table.

There is, therefore, a critical need across the economy for organizations to adopt AI tools and implement corresponding organizational changes. Almost no organization will be untouched. Nonprofits, government organizations, hospitals, and law firms could win some of the larger gains, but (paradoxically) those types of organizations are less likely to implement changes. They can coast on a lot of goodwill and they typically face lower levels of competitive pressure that would otherwise force change.

The technology is advancing rapidly and every industry needs it at the same time. This means that the demand curves for implementers will shift up simultaneously. As a result, the price for this type of skill will increase significantly.

Consulting firms are already scrambling to build the capability to implement AI adoption and organizational change. But the three-part bundle described earlier is hard to assemble inside a consulting firm. Consultants can learn the technology and they can develop frameworks for process analysis. But they typically lack the deep, organization-specific workflow knowledge that only comes from years inside an institution. A McKinsey team can diagnose a problem in six weeks, but the implementer who can move the organization through the valley needs the trust that no external firm can provide.

A limited, rigid supply

The supply of people with the right skills is probably hard to expand. The three domains that our analysis identifies are not easily susceptible to formal training. The technological part can be obtained formally (or through online learning), but the technology moves so fast that people must constantly relearn. It is more likely that people with technology skills will be self-taught. They will keep experimenting and tinkering with the tools for their own purposes, like Maple drawing crime maps on the wall before geocoding software existed.

Understanding the workflow is even harder. It requires people to acquire a deep knowledge of each industry or organization. The knowledge about how a bank compliance process works is full of details and specific rules and routines that are hard to learn from a textbook.

As to politics, being able to identify who will have the motivation and power to block necessary organizational changes, and how to work with those people, is not an easily acquired talent. Likewise, knowing how to build coalitions with those who will benefit from the reorganization can only be learned through deep experience.

We find Jack Maple's example instructive. It is best to find someone down the ranks who understands and cares deeply about the substantive problem, and who understands departmental politics, and then teach them about the technology. That is easier than trying to teach a software engineer about politics and processes.

Regardless, the supply of people of this type is limited and hard to expand. The number of people who combine current technical knowledge, deep workflow knowledge in a specific domain, and the political skill to manage a transformation is small. Training programs can stretch this supply at the margins, but the binding constraint is experiential knowledge, and that takes years to accumulate. No bootcamp can fill the demand gap. Expanding supply probably requires CEOs with the ingenuity of Bill Bratton, who plucked Jack Maple from deep in the ranks of the NYC Transit Police.

AI will not help. The premium here is on skills that AI cannot replicate: tacit knowledge of how organizations work, the ability to build trust, the judgment to know which of a thousand possible reconfigurations to try first.

Demand for implementation skills is large, synchronized across the economy, and growing with every improvement in AI capability. Supply for these skills is small, hard to expand, and dependent on forms of knowledge that take years to develop. The gap between the two is, we believe, the widest of any scarcity described in this book.

The evidence, even at this early stage, supports our view. A 2024 Boston Consulting Group survey of nearly eighteen hundred C-suite executives found that only 25 percent of companies had moved beyond the pilot phase to scaled AI deployment. The remaining 75 percent were stuck in various stages of experimentation. Only 26 percent of companies have developed the capabilities to move beyond proofs of concept and generate tangible value.[120] This is not for lack of ambition or spending. Global corporate investment in AI exceeded $200 billion in 2024.[121] The money is flowing. The transformation is not. The bottleneck is not the technology or the capital. The need is for people who can make the technology work inside complex, messy organizations.

The market implication follows directly from the analysis. The financial returns to those with implementation skills will rise as long as AI capability continues to advance faster than organizational adoption, which is to say, for a long time. This is not a temporary bottleneck that will be resolved as firms learn. It is a structural feature of how organizations work. The complementarities, the fog, and the resistance described in this chapter are permanent properties of complex institutions. AI makes the prize larger; it does not make the obstacles smaller.

Conclusion

People with implementation talent and experience will be in high demand in the next decade. The supply of these people will be small. The work is irreducibly human. It can be placed, firmly and permanently, on the human side of the autonomy threshold.

As a result, most existing companies will fail at adapting their systems to a world of agentic AI. Economist Joseph Schumpeter (1883-1950) put it best: "The problem that is usually being visualized is how capitalism administers existing structures, whereas the relevant problem is how it creates and destroys them."[122] As with most previous disruptions, more newcomers than incumbents will adopt the technology and thrive.

But that is not the whole story. Within every large organization, there are people like Jack Maple. They are buried in the ranks, close to the real work, and frustrated by the gap between what is possible and what is actually done. The organizations that find them, empower them, and let them lead organizational change will be the ones that escape pilot purgatory. The rest will coast on goodwill until the goodwill runs out.

Up to now we have been discussing relationships *within* organizations. We turn next to relationships in the market, between producers and consumers. We shift from supply-side messiness to the value of work that meets the messy demand.

CHAPTER 8

The Humanness Premium

The Demand-Side Threshold

In 1997, IBM's Deep Blue defeated Garry Kasparov, the reigning world chess champion.[123] By 2017, Google DeepMind's AlphaZero had taught itself the game from scratch in four hours and then beat the strongest chess engine in existence without losing a single game.[124]

If the value of chess is based solely on the production of high-quality moves, the market for human play should have collapsed. It did not. Chess has never been so popular. Since AlphaZero killed human chess, viewership, sponsorship, and participation have reached historic highs. Engine-versus-engine tournaments remain a niche curiosity.

The game's value has persisted because of something an AI program cannot supply: the spectacle of humans under pressure. In the February 2026 freestyle chess tournament, Magnus Carlsen was in a losing position against Fabio Caruana after the first two games. Social media was abuzz with the question of whether Magnus could recover from his errors. To the satisfaction of his fans, he did, and won his twenty-first world title in an impressive show of resilience.[125]

A blunder by an AI engine is just a software bug. A blunder by a grandmaster in a world championship is a tragedy that millions will replay and discuss. The product is not chess; the product is the *humanness* of chess. The moves on the board might be identical, whether produced by a human or a machine, but the economic value depends entirely on their origin. This point was illustrated by the Magnus Carlsen/Hans Niemann controversy of 2022–2023, which turned on whether a sequence of moves was generated by a human mind or by a chess engine. If the engine made the moves, they were fraudulent. If a human made the moves, he was a genius. The entire value was determined by who made the moves.

This chapter examines the market-side version of tier three: outputs whose value depends not only on what is produced, but on the human relationship behind it. The value of human origin operates through two distinct channels: instrumental and intrinsic. In the *instrumental* channel, products of human origin carry information that would be lost if a machine produced the same output. In the *intrinsic* channel, people value humanness directly. They want to share a concert hall with a performer, to own what others cannot, and to be served by a person. We study each and then ask what conditions must hold for humanness to command a premium. We also look at how AI can change the size of these markets.

Instrumental value

When a buyer cannot verify quality directly, the seller needs a way to communicate it. A signal, which is an action or attribute that reveals something hidden, does this job. Michael Spence showed that a signal only separates high quality from low quality when the signal is costly to produce. The signal works because it is wasteful (or excessive); the waste is what the buyer is paying for.

A recent *Wall Street Journal* story described the work at Crane Stationery in Latham, New York, where twelve master engravers etch custom designs onto copper plates using presses built before the First World War. The margin of error is less than the width of a human hair. Engravers stand for eight hours a day chasing a standard of perfection that their customers expect. Half of all trainees drop out. The company cut off holiday sales two weeks early this past season because it could not fill orders.[126]

AI could design the layout. A laser already etches the copper plate. But the counter, the thick card stock that pushes paper into the plate to make the impression, must be crafted by hand. Each color has its own plate, and if the alignment of a miniature ballerina is off by a hair's width, the engraver starts the entire process over. A hand-engraved card from Crane costs $15 because a human stood at an antique press for the time it took to get the alignment right. The company's investment in the product is a signal of value for the consumer. It proves that the product's maker values the recipient enough to invest in making something difficult. By contrast, an AI-generated card, however beautiful, cannot carry that signal because the cost of producing it is near zero. Reliable signals require real cost, as we shall discuss in part III.

In a market flooded with synthetic content, products and services with verifiable human origin may increase trust that the product is high quality. In wholesale trade, for instance, 1.6 million sales representatives maintain relationships with buyers. Those buyers could place orders through automated procurement systems, but they choose not to.[127] The informational task can be automated, but not the trust built through personal dealings.

Intrinsic value

The second source of value does not depend on information, but on human preference. The literary scholar René Girard spent decades arguing that human desire is not autonomous but fundamentally mimetic. In other words, we do not want things because of their intrinsic qualities but because other people want them. The child does not want the toy until her sibling picks it up. The handbag is not beautiful until the other woman carries it. Desire, in this view, is not a relationship between a person and an object. It is a relationship between people, conducted through objects.

Research by Alex Imas and Kristóf Madarász shows that people value a good more when they know others are excluded from having it. In standard auction theory, the exclusion of potential buyers should reduce revenue by reducing competition.[128] However, Imas and Madarász demonstrate the opposite: When a seller randomly excludes some potential buyers from the opportunity to purchase an item, the bids among the remaining buyers rise substantially and total revenue increases. The winner pays a premium for the status of possessing what others cannot.

When AI makes a product abundant, it destroys the scarcity that generates mimetic desire. A hand-engraved card is desirable not because it has better quality than an AI-generated one. Its value is affected by limited supply. Human origin is a reliable source of limited supply. Machine origin is not. Recent experiments by Imas and Mandel show this.[129] When subjects bid on physical art prints, human-made work gained 44 percent in value by being exclusive; that is, one copy rather than many. AI-generated work gained less than half that. The mere involvement of AI made the work feel inherently reproducible, regardless of how many copies were said to exist. Scarcity engineered around a machine-made object does not generate the same willingness to pay as scarcity that flows from human origin.

This is the main reason for the value of many luxury brands. A Hermès Birkin bag, the ultimate status symbol in our economy, is handcrafted by a single artisan in France from start to finish. Getting one requires building a history of purchases, by buying other items, such as shoes or scarves, at the same boutique with a particular sales associate, and then waiting for months or years for an offer.[130]

Direct demand for humanness

In many instances, humans directly demand the "human touch." Field experiments find that simply disclosing that the "helper" is a chatbot can reduce purchase rates by 80 percent even when the underlying service is unchanged.[131] Other experiments show that people rate the same quality more negatively when they believe it came from a bot rather than a human.[132] Research suggests that identical products can command a price premium of up to 60 percent when marketed as human-made.[133]

Consider the sommelier. A wine-recommendation app can scan a label, cross-reference it with a database of tens of thousands of reviews, and suggest a pairing in seconds. Vivino, the most popular of these apps, has over fifty million users.[134] The informational task of the sommelier, consisting of matching a wine to a dish, a price point, and a preference, has been automatable for some time. And yet fine restaurants continue to employ sommeliers. The more expensive the restaurant, the more likely it is to have one. The sommelier is not selling information. She is selling a performance: the story about the vineyard, the gesture of pouring the first taste.

These patterns are not confined to retail. Live music is growing at the same time as Spotify and other online music services make recordings essentially free. Recorded music is better produced, better mixed, and available at near-zero marginal cost. Nevertheless, global live music revenue reached roughly $35 billion in 2024 and is projected to nearly double to over $60 billion by 2034.[135] Live Nation, the industry's dominant promoter, reported that 151 million fans attended more than fifty-four thousand events in 2024, and its revenue hit $23 billion.[136] It is not just the famous artists who perform live. Many small venues advertise the presence of a live band. The product is not the music. The product is the human being performing under conditions of uncertainty, in a room with other human beings.

The same pattern runs through the food economy. As industrial food became cheaper and more uniform, a segment of consumers demanded products whose value derives from their origin. Craft breweries in the United States numbered about fifteen hundred at the turn of the century and fell to a trough of 1,394 by 2005.[137] But by 2023, there were nearly ten thousand, and their retail sales reached $28.8 billion, roughly a quarter of the entire $117 billion US beer market by value. This happened even as total beer consumption stagnated, with overall US beer production and imports falling by 1 percent in 2024.[138] Americans were not drinking more beer, they were paying more for beer that came from a specific place, made by specific people, using a specific process. The cheap and efficient mass-produced product lost market share to the expensive, local one.

Something similar is happening in the Indian handloom sector, where technologies like Internet of Things (IoT) sensors are being deployed to authenticate their work. The technology records the strokes of a loom and proves that a specific artisan produced a specific fabric.[139]

Jobs and earnings: Baumol's cost disease becomes a blessing

Up to this point in the book, we have discussed the autonomy threshold as a supply-side concept. The threshold is the point where the technology can deliver a finished output without human involvement. This chapter adds a demand side. A Crane engraver does clean, verifiable, single-task work, which is among the most threatened by AI, at least according to supply-side logic. However, we need to consider demand-side factors. In the case of Crane's cards, customers care who made them. The sommelier's informational task is automatable, and AI has crossed the supply-side threshold, but demand for human interaction keeps her in the job. A personal trainer must read the client's mood, adjust the program to her tiredness, build a relationship, and keep her motivated over months. The client wants to be trained by a person, not an app. This is messy work that may be resistant to the supply-side pressure.

A job survives as long as either the supply or demand side of the autonomy threshold holds. Combine both sources of protection and you may get employment security. What determines the size of the authenticity economy? As before, we need to think of demand and supply.

Demand for human origin

Three conditions determine whether consumers will pay for human origin. The first is *verifiability.* The customer must be able to confirm, or at least plausibly believe, that a human produced the work. As AI-generated content becomes indistinguishable from human output in many domains, verification becomes harder and therefore more valuable when it can be achieved. Physical products have an advantage. You can touch the engraving, inspect the weave, taste the hand-rolled pasta. Music, books, and other information goods may require external certification such as badges, portfolios, or reputation systems to establish provenance. The harder it is to verify, the more the premium depends on trust and relationship.

The second condition is *perceptible quality.* Provenance commands a premium only if the customer believes the human-made product is different in a way that matters to them. This need not mean objectively superior. The mechanical watch is objectively worse at keeping time than the quartz watch. The *wabi-sabi* ceramic in Japan has deliberate imperfections that make it feel human.[140] In both cases, the customer perceives and values the craftsmanship that the machine-made alternatives lack. When the customer cannot perceive a difference, provenance does not generate a premium.

The third condition, for intrinsic or instrumental reasons, is *status or expressive value.* Human origin commands a premium when it allows consumers to express taste, care, status, identity, or belonging. This is the case in markets such as luxury goods, high-end stationery, handmade furniture, artisanal food, and live performances. In these markets, the fact that a person made or delivered the good is part of what the buyer is purchasing. As people get richer, they do not just spend more money; they spend a *rising share* of their income on luxury goods. They do not just eat out more often; they eat at restaurants with more staff per table, where meals take longer and where service is more attentive. When they exercise, they hire personal trainers. Being served by a person is part of what they are buying, which is a direct expression of this income effect. As AI-driven productivity growth makes the economy richer, the resulting income gains are likely to push consumers to spend *more* on human-intensive services, not less. [141]

When any of these conditions fail, demand may suddenly collapse. The NFT art market is an example of this. In March 2021, Christie's sold a digital collage by the artist Beeple for $69.3 million, making him the third-most-expensive living artist at the time.[142] The work existed as a JPEG. What the buyer received was a nonfungible token, a unique entry on a blockchain certifying that he, and only he, owned this particular file. The promise was extraordinary. The blockchain, its advocates claimed, had solved the oldest problem in the art market: provenance. An NFT could prove ownership, certify authenticity, and manufacture scarcity for an object that was otherwise infinitely reproducible. Walter Benjamin, the cultural theorist who argued in 1935 that mechanical reproduction would destroy the "aura" of artwork (its unique presence in time and space), had apparently been overturned by cryptography.[143] The blockchain could create "aura" from code. For about eighteen months, the market believed it. NFT art trading volume hit $2.9 billion in 2021.[144] Then it fell apart. By 2024, trading volume had dropped 93 percent from its peak. The share of collectors buying digital art fell from 17 percent in 2022 to 9 percent in 2024. The number of active traders collapsed from over 529,000 in 2022 to under twenty thousand by early 2025.[145] Christie's quietly closed its digital art department.

Supply of human origin

On the supply side, the binding constraint is that authentic human work cannot be scaled. A live concert cannot be replicated by adding servers. An engraved card requires a human at the press. A Hermès Birkin bag is handcrafted by a single artisan from start to finish. The skill must be genuinely difficult to acquire. Crane's engravers train for at least a year on a single color before advancing to complex designs. If the skill were easy, everyone would do it and the premium would collapse.

William Baumol noted in the 1960s that a string quartet still takes four musicians and the same amount of time to play as in Mozart's time. Quartets have enjoyed no productivity gains. Yet musicians today earn much more than musicians in Mozart's time, because, given the growth in compensation in other sectors, a stagnant salary would deter people from becoming musicians.[146] If the job of hairdresser, whose productivity has not increased in a hundred years, were paid as much as a century ago, no one would do the job.

Baumol's general observation is that wages in the sectors that experience no automation, such as the performing arts, education, or healthcare, must rise as much as wages in sectors where productivity is exploding. Otherwise it would be difficult or impossible to keep attracting workers. Hence, sectors with little or no productivity growth see their costs and wages rise relative to the rest of the economy.

In traditional economics, this was treated as a problem, an unfortunate fact of reduced aggregate productivity growth. But in a world where cognition is abundant, the logic has another, more positive, interpretation.

When AI automates commodity production, prices in that sector fall. Falling prices raise real income. Rising real income shifts spending toward sectors with higher income elasticity; that is, toward goods and services whose demand grows faster than income.[147] Research by Comin, Lashkari, and Mestieri shows that income effects account for over seventy-five percent of observed sectoral reallocation across the twentieth century. Price effects, the standard story that cheap automated output frees up budgets for other things, explain only about a quarter of it. Based on this result, Alex Imas argues that the dominant force in AI-driven structural change will not be automation that pushes workers out of cheap sectors. Rather, rising incomes will pull spending toward sectors where human involvement is part of the value.

Baumol's cost disease then reinforces this shift. The sectors Baumol treated as "diseased," the ones where productivity cannot explode because the work is irreducibly tied to human time and presence, see their relative price rise exactly when demand is already moving toward them. A live performance cannot be diluted by automation. An engraved card requires a human at the press. The zero-productivity growth that Baumol lamented becomes an asset because it guarantees the product cannot be replicated at zero marginal cost. Since higher-income households spend a larger share of their consumption on labor-intensive goods and services, growth itself tilts demand toward sectors with higher labor content.

When considered together, falling prices, the income effect, and the Baumol amplifier predict that the share of economic activity devoted to human-presence and human origin, both products and services, will *expand* rather than contract, as AI makes cognitive work cheap.[148]

Three tiers, three economies

The result is that there are two polar-opposite economies running in parallel, with a wide middle. In the algorithmic economy (the old tier-one jobs), AI and automated logistics produce good quality output at near-zero marginal cost. This economy will drive down the price of cognitive commodities such as reports, code, translations, designs, marketing copy, and standard legal documents. The value in this economy is captured by the asset owners, the people who own and control the distribution or the platform, not by the people who produce the content.

On the messy side of the economy (tier-three jobs that are permanently above the autonomy threshold), value comes from friction and imperfection. These jobs are intrinsically human activities that preserve value at any level of automation. This includes those involving authority and implementation (chapters 6 and 7), and authenticity and provenance.

In between, we have a wide set of tier-two jobs, the bundled professional work described in chapters 3 through 5. Examples include the doctor who uses AI diagnostics but still holds the patient's hand; the entrepreneur who builds a company with AI handling her weak links; the restaurant manager who uses AI scheduling but manages the kitchen crew. These workers are neither replaced by the algorithm nor protected by provenance. They survive because their jobs bundle cognitive tasks with social, physical, and relational ones that AI cannot perform. The bundle holds together. Most careers will sit here in tier two; not in the algorithmic economy, not in the provenance economy, but in the messy middle working with AI.

The three tiers describe what happens to jobs. But jobs exist inside organizations, and organizations were designed for a world where every bundle was purely human. Hiring systems assumed human-produced signals. Monitoring systems assumed output was produced at the speed of a human. Training systems assumed human apprentices doing human routine work. AI changes every one of these assumptions simultaneously.

Part III examines what breaks and what must be rebuilt. The complementarity logic from chapter 7 applies to this question. Organizational systems are clusters of interlocking features, so when one feature changes (e.g., production technology), the others must change together. The result is that organizations must redesign at least four core systems: screening, verification, performance standards, and careers and training. Each redesign also has direct consequences for the workers inside the system.

Part III

Organizing the Human-AI Bundle

Parts I and II asked which jobs survive AI. But jobs do not exist in a vacuum. They exist inside organizations that were designed with multiple human-centric systems for hiring, monitoring, training, and other functions. Layering AI on top of these systems will not work. The systems themselves must change. That is the focus of part III: how organizations must redesign screening, verification, standards, and training when every surviving job becomes a human-AI bundle, and what that redesign means for the workers inside them.

CHAPTER 9

Compressed Signals

Redesigning Screening

A $5.3 million seed round shouldn't have gone to a college dropout who cheated in an interview. But that's precisely what happened to Roy Lee in 2025.[149]

Lee, a twenty-one-year-old former computer science student at Columbia University, built Interview Coder, software that gives real-time answers onscreen during technical interviews. Think of it as a silent genius crouched under the desk. He used it to ace an Amazon coding interview, got exposed, and was promptly suspended. Instead of apologizing, he rebranded.

Within a month, Cluely was born: "AI for cheating on everything." Lee's launch video showed him wearing smart glasses during a dinner date. Cluely fed him answers to his date's questions in real time, making him sound knowledgeable and charming. Within months, Lee had $5.3 million in seed funding and thousands of subscribers paying to sound brilliant.[150]

Cluely's story reveals a structural problem for organizations. AI has improved productivity and empowered workers to do things they couldn't do before, but it also changes the way people produce information about themselves. When everyone can use Cluely to answer tough interview questions, the ways candidates demonstrate that they are competent and worth hiring are fundamentally altered.

This chapter is about how organizations can screen credibly when AI makes it easy for anyone to fake competence. How can a firm know that a candidate deserves the job or a promotion? To answer that, we need to understand why people rely on signals in the first place, how others interpret those signals, and how AI upends both.

The signals we send

Employers cannot see your most important qualities. They cannot see your intelligence, your judgment under pressure, how hard you work without supervision, or whether you are genuine and enthusiastic about your project. They only see surface-level signs: your résumé, letters of recommendation, and the hours you spend in your office. These appearances are the signals, and their value depends on how closely they relate to your real qualities.

Signals simplify the employer's job. Employers do not have to spend days figuring out your true intelligence. Instead, all they have to do is see which school you graduated from. A well-written cover letter demonstrates your passion for the job and why you are a perfect fit. Working long hours proves that you dedicate yourself. Each sign condenses complex and hard-to-see information into concise, easy-to-understand content.

Manipulated signals

Here's the problem: Some signals genuinely reflect your quality, but others can be manipulated. We can choose which signals to send, and we can distort them. When we send manipulated signals that exaggerate our abilities, those signals create a gap between genuine quality and perceived quality.

At prestigious law firms, the partners have no real idea who the best associates are. Talent is invisible, and effort is noisy and hard to verify. All the partners can see are billable hours, emails, and the hum of late-night printers. So the associates compete in one type of signal the firm can measure: hours.

A survey at two major firms found the median associate logged about two hundred hours a month, with the top quarter reaching 220 hours. Two-thirds said they would happily reduce their hours by five percent, yet few dared. Everyone understood that working less might look like caring less. So the office became a kind of theater. Associates scheduled emails for 3 a.m. They left jackets draped on chairs overnight to suggest they had stepped away only briefly.[151]

The rat race

Manipulating signals to make yourself look good can be exhausting. It gets worse when the goal is not to look good in isolation but to look good relative to everyone else. If you were the only associate working late, staying until 11 p.m. would be a costly but powerful signal. But when everyone stays until 11 p.m., that merely becomes the baseline. The signal's value depends not just on what *you* do, but on what everyone else is doing. Because you cannot control what others do, you are forced to keep up or risk falling behind.

This is why signaling so often turns into a rat race. Each person inflates their signals a little, which forces everyone else to inflate their signals a little more, which resets the standard upward. Dating markets are a good example. On Match.com, people exaggerate their qualities not out of overconfidence but because failing to do so puts them at a disadvantage. They may not even get a first date. People who claim to be "athletic and toned" are closer to "a few extra pounds." Because exaggeration has become the norm, being honest and stating that you have a few extra pounds makes you sound significantly overweight. Honesty becomes self-defeating, not because honesty is bad, but because everyone else's inflation has shifted the scale. Daters have to exaggerate just to stay in the game.[152]

Under Stalin, losing the rat race of competitive signaling could kill you. At a Communist Party conference in the mid-1930s, delegates rose to applaud the "great leader." Once it started, the applause refused to end. Five minutes passed, then ten. The delegates felt the pain of aching palms, but no one dared to be the first to stop. After eleven minutes, the director of a local paper factory finally stopped. That night, he was arrested.[153]

The law firm, the dating profile, and the applause for Stalin all share a common structure. There is a gap between who you really are and what other people can observe. In many situations, it is the perception that gets rewarded, not the reality underneath. And so enormous energy goes into performing qualities that people may already possess. People put in effort not just to do the work but to perform in the workplace theater, and the performance of competitive signaling never ends.

The economics of showing off

So far, we have looked at signaling from the sender's side: the associates staying late, the daters polishing a profile, the delegates clapping until their hands burn. There is another side to the story. Employers, clients, and leaders also have a problem. They must figure out who is actually competent and a good fit. Jobs, promotions, and attention are scarce, so allocating them well requires information that is expensive to obtain.

Some qualities can be tested directly. If you want to know whether someone can play the piano, ask them to play. If you want to know whether an artist can draw, hand them a pencil. These are skills, easy to observe and easy to verify. The demonstration is the proof, and there is no need for strategic signaling.

But many of the qualities that matter most in professional life are not like that. Loyalty, diligence, and commitment are hidden. You cannot ask someone to "perform" loyalty the way you can ask them to play a piece of music. You cannot verify resilience in a thirty-minute interview. For qualities like these, direct demonstration is impossible. However, it might be possible to signal these qualities through a willingness to bear a cost that someone without the quality would not be willing to bear.

Sima Qian's Shi Ji records a famous case of this from the late sixth century BC. Rangju, a low-born officer in the state of Qi, was appointed to command the army against invading states. Nobody doubted Rangju's knowledge of military strategy, but no one knew whether he had the resolve to lead an army that did not yet respect him.[154]

Rangju asked the duke to send his favored courtier, Zhuang Gu, to oversee the campaign. Zhuang Gu showed up late to a military parade. Rangju had him executed on the spot. When the duke's messenger arrived to protest, galloping through the camp, Rangju executed the messenger's attendant.

This was not a demonstration of ability. It was a demonstration of his willingness to bear an enormous cost to ensure discipline. Offending the duke could have ended Rangju's career or even his life. Only someone genuinely committed to military discipline would accept that risk. After the executions, nobody in the army doubted him again. He went on to win many victories.

Michael Spence, who won a Nobel Prize for formalizing this logic, showed that the same principle governs much of economic life.[155] Consider the MBA. Graduates command higher salaries immediately upon graduation, often before having had time to apply what they learned. This seems puzzling until you see it through Spence's lens. Beyond the knowledge acquired, the degree proves that you are the type of person who thinks the pain is worth it. A strong student expects to gain more from the MBA through a steeper career trajectory. Therefore, the tuition and the two years of forgone earnings are a worthy investment. A weaker student faces the same price tag but with a lower expected returns, so they rationally opt out. The degree separates students by creating a cost that is worth bearing only for those with the underlying quality.

The same logic shows up in business. An entrepreneur who is confident her venture will succeed is willing to rent the expensive office downtown because the clients it attracts will more than cover the cost. One who doubts his prospects sees the same rent as money thrown away. The office signals quality because the investment only makes sense if the business is genuinely strong.

Once you see this logic, you start noticing it everywhere. Accepting an unpaid internship is costly, which is why it signals commitment. The sacrifice pays off only if you intend to build a serious career in the field. Expensive handbags and fragile watches are not primarily bought for comfort; rather, they are bought to show that you can afford a cost that would be ruinous for most other people.

A signal carries information only when it fulfills two conditions. First, it must be costly to produce. Second, its cost must correlate with the quality being signaled. If a degree were easy, it would prove nothing about the talent or work ethic of the candidate. If a marathon were easy, finishing one would not impress anyone. If you remove the cost, you also remove the message.

When AI compresses signals

AI changes how workers send signals and how organizations interpret them. Cluely, the software that feeds candidates answers during live interviews, is perhaps an extreme example. A less extreme one is how AI is reshaping one of the most basic signals in hiring: the cover letter.

Researchers have used data from online labor market platforms to study how AI changes the way cover letters are created and used.[156] Traditionally on these platforms, an employer posts a job. Within a day, thirty or forty applications pour in. Most are generic, but a few applicants have read the job description carefully. They mention specific requirements and explain why their experience fits the project. Their cover letters stand out.

Before AI, the quality of a cover letter was a useful signal. Workers who spent more time tailoring their applications were more likely to get hired and more likely to do good work once hired. The cover letter was a valid screening tool.

The cover letter was never a perfect filter, especially for non-native English speakers. A software developer in Bangalore or a designer in São Paulo might have been exactly the right person for the job, but if their cover letters were poorly written, the employer probably never got far enough to find out.

When freelancing platforms introduced AI writing tools, these workers were among the first to benefit. One click transformed a rough draft into a tailor-made cover letter. Those who used AI assistance were initially more likely to be hired. But the equilibrium effect made matters more complicated. Suddenly, everyone had access to the tool. The worst writers gained the most because they had the most room to improve, but the naturally good writers no longer had the same advantage. The gap between the careful letter and the lazy one almost disappeared. Better workers were hired less often, and the wage premium they previously commanded vanished.

In the freelancing market, and elsewhere, organizations suffered the consequences of the cover letter's loss of value: They often hired the wrong worker. A tool meant to help everyone communicate better made the market worse at matching workers to jobs.

This is what Spence's model predicts. A signal only works when its cost correlates with the quality being signaled. A cover letter worked because writing a good one required a combination of ability and effort. AI is upending the screening process, and organizations must rebuild it.

How organizations must adapt the screening process

When AI compresses one type of signal, screening shifts to whatever signals still convey information. Organizations are shifting their attention toward signals that AI has not compressed, such as past work experience and platform scores that aggregate many years of feedback. This is not a disastrous outcome for established workers, but it is a disaster for newcomers who have not yet had time to build a history.

The same pattern can be observed beyond freelancing. In China, where postgraduate degrees have become easier to obtain and embellish due to the expansion of higher education, the undergraduate degree has regained importance among employers. The reason is that it rests on the *gaokao,* a national entrance exam that is hard to game. As the signals that come later in a career grow noisier, the ones that come earlier and are backed by transparent competition become more informative.

Also, organizations are increasing the weight of in-person and real-time interaction. In some firms, interviews are much longer and include exams, case discussions, and real projects, precisely because they are harder to game. The cost of showing up in person and the awkwardness of being rejected are what make it informative. Even being willing to attend a long unstructured interview becomes, in Spence's terms, a signal whose cost still correlates with quality.

Organizations are also increasing the requirement for candidates to show significant upfront work, another signal that separates the lazy from the committed.

How can individuals adapt? Here's what we tell our students: Rather than using AI to reduce the time you spend on a mediocre project, use AI to produce substantially better projects. Use the time AI saves you to do high-quality work that AI cannot reach. The competitive advantage is not the tool, because everyone has it. The advantage is what you can do with the time it saves you. This is the paradox of signal compression: AI destroys the value of the signals it can easily produce, but in doing so, it raises the value of everything it cannot do.

While writing this chapter during Chinese New Year, Jin received a pleasant phone call from his friend Evan Guo, the CEO of Zhaopin.com, simply to exchange greetings. Evan mentioned that AI now allows written greeting messages to scale infinitely, effectively rendering them meaningless. By comparison, a phone call requires time, and time remains scarce. Because it cannot be effortlessly automated and easily scaled, committing your time to call someone shows genuine care and commitment. For organizations, this logic extends to their client relationships, their hiring processes, and every context where the cost of attention is the signal that matters.

AI has not changed what employers, clients, and collaborators ultimately want to know about you. They still want to know whether you are competent, whether you can be trusted, and whether your judgment is sound. What *has* changed is the channel. The polished résumé and the eloquent cover letter can no longer serve as credible signals. The new screening tools that have taken their place are older and simpler: verifiable track records and real-time interaction.

Organizations that redesign their screening around these channels will hire better employees. Workers who understand this logic will invest accordingly.

CHAPTER 10

Workslop

The Verification Crisis

Chapter 9 was about selecting the right people. This chapter is about what happens inside the organization after hiring; specifically, how to maintain quality when AI severs the link between surface and substance.

In late 2023, Amazon.com was flooded with wild-mushroom foraging guides. These fluently written guides had nice covers and were incredibly detailed in taxonomy, as if the authors had done years of muddy fieldwork. However, the New York Mycological Society quickly issued an urgent public warning: "Please only buy books by known authors and foragers. It can literally mean life or death."[157]

The Mycological Society was not being dramatic. After reviewing many books on Amazon, experienced mycologists found dangerous misidentifications. Poisonous fungi were mislabeled as edible and descriptions were so vague that they confused the deadly death cap with common field mushrooms. These "authors" did not exist, and their credentials were made up. The books were generated entirely by prompting a large language model and then uploading the unedited output to Amazon's self-publishing platform. From prompting to live listing, the entire process took just one afternoon.

Before the advent of AI, authors with specialized knowledge of mushrooms were needed to write mushroom foraging guides. The friction of the writing process itself acted as a natural filter. In-depth knowledge of the field was required to produce credible and accurate guidance. With AI, someone with no knowledge of mycology can write a book that looks no different from one written by an expert. The cover looks nice and the tone is authoritative, but the advice inside might kill you.

The mushroom guide example is an extreme case, but the underlying problems are much more widespread. Cheap production, costly verification, and weak incentives to check errors have become common features in knowledge work. Consider a university professor whose career depends on the number of publications he produces. AI can now draft a passable literature review in hours. He possesses the expertise to verify every claim, but he also has three other papers in progress, an overflowing inbox, and a grant deadline on Friday. He takes a quick look at the literature review and moves on to other tasks. The professor is not a scammer selling fake mushroom books, but he is responding to the exact same incentive.

The previous chapter showed how AI compresses the signals that organizations have traditionally used to screen for quality, making it difficult to spot talent. This chapter is about whether an employee's work can be trusted. AI makes trust harder to sustain in knowledge work because of a fundamental problem of organizational life.

Organizations face an important new phenomenon called *workslop*. This term refers to output that mimics the form of real work without embodying its substance. With AI, the cost of producing passable output has collapsed. An analyst can now prompt an AI, get a polished-looking draft in minutes, and leave early. What once required an evening of focused work now takes five minutes of typing.

The fundamental incentive problem

The problem of monitoring and creating incentives for effort is so universal that every student has lived a version of it. It's called the group project. Four people, one shared grade. Brad contributes a single slide with three generic bullet points. Jennifer disappears entirely. Mike promises he is "almost done." And you find yourself stitching everything together alone at 3 a.m.

It is comforting to blame bad character; in other words, to say that Brad is lazy and Jennifer does not care. But deeper forces drive the failure of group projects. First, the group's collective needs are not necessarily aligned with what each member wants. Each hour you spend on the project is an hour you cannot spend doing something else. This creates an obvious temptation: let others do the work while you enjoy the shared result. When effort is costly to you and the rewards are shared collectively, the rational response is to let others carry the load. After all, everyone gets the exact same grade whether they contribute or not. This is called the *free-rider problem,* a classic reason for the breakdown of team production.[158]

Second, individual effort is incredibly hard to observe. The most valuable inputs in knowledge work, such as thinking, judgment, and care, are not easily recorded and measured. The professor cannot see who stayed up until 3 a.m. wrestling with the ideas.

When unobservable effort meets misaligned preferences, it

creates what economists call the *moral hazard problem.*[159] The junior analyst knows whether she stress-tested her financial assumptions, but her manager does not. The consultant knows whether he verified the data, but the client sees only the polished slide deck. In each case, the person doing the work knows something about the quality of her effort, but no one else can observe her labor. Furthermore, because the person doing the work only reaps a small fraction of the final benefit, she faces a temptation to cut corners. It is not about figuring out your underlying ability (which we studied in the previous chapter). It's about ensuring you do the work.

Similar to the signal manipulation problem mentioned in the previous chapter, moral hazard is rooted in *asymmetric information,* which pertains to situations in which one side knows something the other does not. Signal manipulation is about figuring out who someone is, but moral hazard is about getting them to do the right thing.[160]

The moral hazard problem turns out to be a central issue for all organizations. Some economists believe that it is the basic reason why organizations exist. If individual effort and the true quality of output were easy to measure, we would not need companies or managers; we could simply buy and sell every task on the open market. But because the actual quality of work is hidden from the outside, firms arise to manage the gap. They use a combination of tools, including incentives, monitoring, and culture, to get people to do what they cannot be formally required to do. These tools help, but they are also costly and limited.

Why moral hazard is hard to fix

A natural solution to the moral hazard problem is to have somebody monitor the effort. Economist Steven Cheung illustrated this with the story of riverboat haulers in pre-Communist China. These workers dragged boats upstream using heavy ropes. The work was backbreaking, but it was also easy to pretend to work hard and let the others do the pulling. The haulers dealt with this problem by pooling their money to hire a monitor who carried a whip. His job was to strike any hauler who appeared to be slacking.[161] It seems strange that workers would voluntarily pay someone to punish them, but without the whip, the temptation to shirk was strong. Workers worried that the boat might not move at all.

The whip is an extreme form of dealing with the moral hazard problem, but its spirit carries over to modern organizations, even if the tools are somewhat more civilized. Managers are the monitors. Their "whips" are bonuses, promotions, performance reviews, and the threat of termination.

Paying managers to watch workers can be very expensive. The manager must spend time evaluating rather than producing. Furthermore, it is far more challenging to monitor knowledge work than physical work. You cannot physically watch what someone is thinking. And who will monitor the monitor?

Because paying managers to watch everything is too expensive and often impossible, organizations turn to a second tool: explicit pay for performance. Instead of trying to watch the invisible effort, firms measure the visible output. They track the deliverables, score performance, and tie the worker's pay directly to the numbers.

But even if they try their best, firms can only measure a proxy for the performance they truly care about. And because workers are paid for measured performance rather than for true performance, they direct their effort to make the measured performance look good rather than to advance the true interests of the firm.

Economists Bengt Holmström and Paul Milgrom, both Nobel Prize winners, took this observation seriously and coined what they called the *multitasking problem*.[162] A professor is supposed to both teach and conduct research. Research is cleanly measured by publications, but teaching is measured by course evaluations that most faculty consider noisy and unreliable. Predictably, the professor focuses effort on publishing, which means that teaching can suffer. This is not because the professor does not value students, but because the reward system does a poor job of valuing them.

Even worse, employees can game the system and cheat. When Wells Fargo wanted its employees to sell more products, it measured success by counting accounts. Employees responded by opening millions of fake ones.[163] When New York and Pennsylvania published hospital mortality rates to help patients choose better surgeons, surgeons responded by avoiding high-risk patients. A difficult case might save a life, but it could also add a death to the surgeon's record.[164] In each case, the metric looked good, but the underlying goal suffered.

Even if there is no gaming and the performance measures are reasonably accurate, monitoring and explicit incentives only provide baseline performance. The hauler pulls hard enough to avoid the whip but no harder. Professors give lectures but they do not stay after class to help struggling students. People do the bare minimum and little more.

What organizations actually need is for people to *do the right thing,* not just follow the rules. And doing the right thing is often costly. It means speaking up when staying quiet is easier and slowing things down when everyone wants to move fast.

Consider Toyota's *andon* (meaning lantern or signal) cord, which is a rope that runs along the assembly line and gives any worker the power to stop the entire production line if they spot a defect. Pulling it costs thousands of dollars a minute. No explicit incentive scheme would tell a worker to do that. If anything, a system that rewards output would punish them for it. But Toyota workers pull the cord anyway because the culture prioritizes quality and they trust the company's promise to back them up rather than punish them.[165]

Economists have studied company culture and trust through what they call a "relational contract," an unwritten deal enforced by the value of the relationship itself.[166] I help you now because I expect you to help me later. I will do careful work today because my ongoing reputation in this team depends on it tomorrow. Neither side can enforce these deals in court. They hold because the relationship is worth more than any single act of defection. Organizations are essentially using the future to motivate effort in the present.

In addition, trust radically lowers the cost of management. A manager who trusts her team does not need to check every report line by line. Managers can empower their subordinates to take on promising but risky projects without worrying about failure. Trust efficiently substitutes for distorted metrics and expensive monitoring. When trust is high, organizations can obtain excellence without the high cost of verification and monitoring.

This type of relational contract requires mutual belief that is anchored in a shared future. Workers need to trust that firms will keep their promises, and firms need to trust that workers are doing the right thing without needing to be watched. When that future is in doubt, trust is difficult to sustain. That is the limit of relational contracts.

The breakdown in verification mechanisms

AI poses a serious challenge to the existing way organizations manage workers' behavior. It changes both how work is produced and the difficulty of checking it. Many monitoring, verification, and accountability systems that organizations built over decades have become obsolete.

On the one hand, producing workslop is essentially free. On the other hand, checking and verifying are hard and getting harder. Humans filter information, which means we are prone to missing important details. A classic demonstration makes the point. Read the sentence below and count the number of Fs.

> FINISHED FILES ARE THE RESULT OF YEARS OF SCIENTIFIC STUDY COMBINED WITH THE EXPERIENCE OF YEARS.

Most people confidently answer three. The correct number is six. The Fs in each "OF" slip by because we rely on shortcuts.

Most of the time, shortcuts can be helpful. But when AI severs the link between substance and surface quality, skimming fails.

This creates the verifier's curse. A junior associate can prompt an AI at 6 p.m., receive a fluent, confident draft, and head home. To do her job properly, the reviewing manager can no longer casually skim the document. She must read closely, check assumptions, trace arguments, and sometimes redo the analysis herself. Because AI-generated drafts are often longer, smoother, and more assertive, they are increasingly costly to verify. The real work moves to the more senior and more expensive person.

To make things worse, verification is itself subject to moral hazard. Careful checking is costly, but thorough scrutiny and blind rubber-stamping look identical from the outside. When effort cannot be easily observed and there's little reward for checking, the temptation to skim increases exactly when rigorous scrutiny becomes more important.

In 2025, it was revealed that a government-commissioned report produced by the major consulting firm Deloitte Australia contained fake citations and references to academic papers that had never been written. The report followed the familiar "strategy deck" framework, with confident headers, data-driven language, and clean layouts. Those visual details effectively masked the fact that the underlying evidence was hallucinated. The output looked like a standard professional deliverable, so it bypassed human scrutiny despite Deloitte's institutional quality controls and global brand.[167]

Technology can assist verification, but automated tools often worsen the verifier's moral hazard. In collaborative software development on GitHub, many projects rely on continuous-integration bots to run automated tests before human review. These tools catch obvious bugs and speed up evaluation.[168]

As automated checks improve, evaluators lean on them

more. Passing the checklist becomes a substitute for understanding the work. If automated tests pass, broader questions regarding architecture, maintainability, or subtle risks receive significantly less human scrutiny, especially when there is a tight deadline. Over time, however, projects accumulate serious unresolved problems even as the evaluation process appears highly efficient.

Furthermore, when managers use AI to verify work, participants inevitably use AI to game the system. One example recently emerged in academia. Investigators discovered numerous research papers uploaded to the arXiv preprint repository that contained hidden, white-on-white text designed to manipulate the review process. Knowing that overwhelmed peer reviewers were using AI tools to draft referee reports, authors embedded invisible instructions that requested the AI to return only positive feedback. The verification system was thus turned entirely against itself. Reviewers used AI to avoid the cost of careful reading and authors manipulated the AI to exploit that exact shortcut.[169]

When AI lowers the cost of production and manipulation while simultaneously raising the cost of verification, the gap between what organizations believe is happening and reality grows wider. AI creates a twin crisis. On the worker's side, effort is easily faked and concealed. On the manager's side, verification becomes more demanding. The bundle of trust, monitoring, and incentives that held organizations together is strained, so the organization must redesign it.

This is why trust becomes more important. Trust is essential for producing exceptional outcomes, of course, but it is also vital for defending baseline performance. When monitoring and checking get overwhelmed and when automated checks are exploited, quality assurance must shift from the work itself to the person who creates the work. The question, then, is how organizations can build and maintain trust.

How organizations must redesign verification

If the old systems are broken, what replaces them? When checking each output individually is too costly, and when automated checks are gamed, the organization must shift from verifying work to verifying workers; that is, verification moves from output-level monitoring to relationship-level trust.

The most direct response is to shrink the unit of trust. Four people who have worked as a team for two years can rely on relational contracts. Their shared future is valuable enough that no one has an incentive to submit workslop. A team of forty strangers cannot rely on relational contracts. The implication is that organizations will move toward smaller, more stable working units when trust substitutes for checking.

Small enough teams are not always possible, so organizations must also invest in a transparency infrastructure. This means requiring workers to show their process, not just their output. This can be demonstrated with annotated sources, documented assumptions, and audit trails of how AI was used. The BBVA case from chapter 7 illustrates this. The company's autonomy-responsibility binomial gave employees freedom to experiment with AI, but it made them responsible for outputs, enforced with a signed procedure document for liability. This is an example of redesigned verification. BBVA did not check every output, but it created institutional structures that made careless production visible and costly.

Finally, organizations must redesign the *incentives* for verification. The verifier's curse arises because thorough scrutiny and blind rubber-stamping look identical from the outside. Organizations can reward careful checking and high-quality output by offering workers recognition (e.g., awards) and career advancement opportunities. They can also hold workers explicitly accountable for what passes review. These approaches may partially counteract the verification aspects of moral hazard.

For the individual, the implications follow directly from the organizational logic. Standard career advice for young professionals has been to network broadly. Sociologist Mark Granovetter showed that *weak ties,* such as casual acquaintances, former colleagues, and friends of friends, are surprisingly valuable for finding new opportunities. Because they move in different circles, people you barely know can connect you to information and job openings that your immediate network cannot.[170] Weak ties are engines of discovery, but they cannot vouch for the actual quality of your work. Casual acquaintances who pass along your name are not staking their own reputation on your competence.

What really gets you through the door and helps you get promoted is a small group of people who have observed your work repeatedly and closely in real-world settings. These deep relationships are based on trust. They need to be built slowly through shared projects and verified outcomes. Since both parties want to continue working together, and because the future value of the relationship is significant, neither party has an incentive to cut corners. You won't hand your manager an unchecked AI draft when your promotion depends on her. And after you have delivered reliable work for two consecutive years, your manager probably won't micromanage every one of your citations. In this case, the relationship replaces verification. It shows that when trust is established, the person who is thoroughly familiar with your work can vouch for you to others. That endorsement will carry weight precisely because the endorser's own reputation is also at stake.

The mid-level relationships are more problematic. They are too familiar to provide the unexpected information that weak ties offer, and they are too shallow to build the trust necessary for deep ties. They are time-consuming, potentially crowding out more valuable connections.

Therefore, a practical approach is to use a barbell strategy in your professional network. Maintain a wide range of weak ties, but focus your real, time-intensive investment on a very small number of deep relationships with people who intimately understand the substance of your work.

How do you gain trust during the first few projects, before you have any track record? Front-load. Proactively take on the most challenging and the most unglamorous verification tasks early on. Manually check data. Trace citations to their original sources. Walk your manager through your reasoning step by step. This is the slow, costly work that most people will try to skip.

By investing heavily in building trust upfront, you accumulate what amounts to reputational collateral. You expect to be rewarded later, with autonomy, better assignments, and career advancement. Those future returns will be conditional on your continued work ethic. The more you have invested, the more you stand to lose by cutting corners. This changes how others see you. A manager who knows you have put in months of meticulous work also knows it would be irrational for you to throw away that trust with a careless AI shortcut. Your heavy investment makes it safe for the manager to trust you. That is how you solve the cold-start problem.

CHAPTER 11

The Mediocrity Trap

Allocating Effort Where It Pays

In 2022, at the Colorado State Fair, Jason Allen won a prize with his AI-generated artwork. When he was awarded the prize, he made the claim: "AI won and humans lost."[171]

When the news was released, it created outrage. One Twitter user commented, "Ten years ago when I was telling people I switched to doing digital art, this is what they assumed I did, just pressed a button and generated art. . . . Now I guess they are giving awards for doing just that."

The truth is far more complicated. Jason Allen wasn't just pressing one button. The project took one hundred hours, most of which he spent struggling with the model output. He spent eighty hours on prompting, sifting through hundreds of "not-at-all-satisfactory" samples, and finally achieved that unique, surreal lighting effect. Once the figure was dressed in flowing robes, he imported the work into Photoshop for a final twenty hours of manual refinement. It was a meticulous process of adjusting shadows and fixing details.

Allen's story forces us to confront a new organizational reality. We already know that AI makes it easier to get started, but its impact on finishing work is less clear. When a machine can give you a piece that is 90 percent perfect in ten seconds, where does the motivation to spend ten hours honing the remaining 10 percent come from? The real question may not be how to complete that last 10 percent, but whether that 10 percent is worth the effort.

The previous two chapters have shown that organizations are getting worse at reading the quality of work and verifying results. The signal is compressed. The cost of verification is high. This chapter asks the following questions: How must organizations set standards and evaluate output when AI changes the shape of the production function? And, given those changing standards, where should individuals really spend their time?

The price of perfection

Long before machines could paint, humans were dealing with the problem of perfectionism. Great work has never depended on talent alone; it depends on how much discomfort and delay someone is willing to endure for a slightly better result. If masterpieces have a secret fuel, it is obsession.

Stanley Kubrick was the high priest of perfectionism. During the filming of *Eyes Wide Shut,* he demanded take after take to perfect even the simplest gestures. Tom Cruise described walking through a doorway dozens of times until Kubrick felt the moment was right. The production stretched to forty-six consecutive weeks, setting a Guinness World Record.[172]

There is a flattering modern name for this: grit. Psychologist Angela Duckworth defines grit as the passion and perseverance to achieve long-term goals. A person with grit refuses to accept the tenth draft because the eleventh might be better. Duckworth argues this quality predicts success better than natural IQ. Grit can lead to great work, but it casts a long shadow.[173]

Even the most determined drive eventually has to end. At a certain point, a difficult question appears: Is this small increment of "better" really worth what it costs?

At Pixar Animation Studios, artists call this trap "the beautifully shaded penny." It describes a situation where an artist pours days of labor into a detail the audience will never see, like a penny on a nightstand in the background. In *Monsters, Inc.,* artists carefully designed individual surface effects for ninety CD cases in a character's apartment. But the CD cases appear for less than three seconds, and only a few of them are visible on screen. The effort was impressive, and it was a good story to tell, but the value to the audience was near zero while the cost to Pixar was real.[174]

The shaded penny problem has many victims. Gustave Flaubert, the French writer, was known to pace his study for hours, reading sentences aloud until the rhythm felt right, sometimes discarding days of work for a single misplaced adjective. In his letters, he complained that countless hours were lost while struggling over "five or six lines." It took him nearly five years to complete *Madame Bovary*.[175] Leonardo da Vinci produced thousands of sketches but fewer than twenty completed paintings because of his exacting standards. In modern times, George R. R. Martin's long-suffering readers may never see the next *Game of Thrones* volume.

To deal with the shaded penny problem, creators have invented diverse methods to force themselves to stop. At Pixar, producer John Walker introduced a system involving popsicle sticks. He attached a row of sticks to a wall, each representing one person-week of labor.

If a director wanted to polish a background detail, Walker pointed to the wall and asked, "Which stick should we use?" More time spent on the background meant less time on the main character. Production now requires a trade-off of finite resources.[176]

Graham Greene found another solution. He divided his works into "novels" and "entertainments." The novels carried profound artistic aspirations whereas the entertainments were thrillers written quickly and without the same crushing pressure. By labeling some of his works as "light," Greene freed himself from the demand that every single project had to justify unlimited effort.[177]

In all of these examples, the rules preventing (or stopping) the unlimited expansion of effort. They enable people to complete their work, continue creating, and devote their energy to other things. But they are also imperfect answers. If we want to think about how to effectively solve this problem, thereby allowing effort to freely expand, we need to systematically consider the general principles involved. This again leads us to economics.

The economics of stopping

Economics is often described as the study of markets, prices, and money. But it is simpler than that. Economics is the language of human decision-making. It examines how people make choices with limited time, attention, and energy, and, just as importantly, when to stop.

For many activities, economics shows us that decisions are not simple binary choices of yes or no. The choices are more complex, such as deciding how much to invest. Many choices are incremental.

We decide not only whether to write, but also whether to revise. The question is not just about whether more work is valuable; we also ask whether another step is worth the cost.

The payoffs to almost any activity are not uniform. They follow the *law of diminishing returns.* According to this principle, early efforts often yield significant gains whereas later efforts yield progressively smaller improvements. In the end, people can spend a lot of time with little to show for it.

Consider a writer who is revising a chapter. The first revision restructures the argument and eliminates weak sections, creating a huge improvement. The second revision catches minor errors and tightens the prose. The improvement is smaller, but still worth the time. By the sixth revision, the gains come from small details that readers will hardly notice. This pattern echoes the 80/20 rule: The last 20 percent of the work often consumes 80 percent of the effort.

Given the law of diminishing returns, the question is when to stop. The rational approach is that we should continue only when the benefit of the next step is greater than the cost. We should stop when the additional benefit is equal to the additional cost. Beyond this point, continuing will result in a loss. The value of our time is greater than the value of the results.

The stopping point is not just a property of the task. It is also a property of what the person could be doing instead. Economists find that older individuals, retirees, and the unemployed systematically spend more time shopping and searching for lower prices. This is not because they benefit more from finding a bargain, though they might. The more likely reason is that they have more time. An hour spent searching for lower prices is not burdensome for a retiree. It is a burden for a busy banker. Both people stop when the benefit of continuing falls below the cost, but that cost is different for each of them.

Diminishing returns in quality do not always mean diminishing returns in value. An Olympic sprinter who shaves 0.02 seconds off her time has made a negligible improvement in absolute performance. But if that 0.02 seconds is the difference between gold and fourth place, the return in value is huge. Similarly, a marketing pitch that is 5 percent more polished than the next firm's pitch wins the entire contract. In these cases, the quality curve is flat but the reward curve rises sharply. This means that rational effort allocation is not simply a matter of examining the quality curve. You also need to know the reward structure. And this, as we will see, is a key issue in the age of AI.

When AI frontloads progress

How does AI affect the decision to continue or stop? The answer lies in how AI reshapes the trajectory of progress.

The new shape of progress

Starting a project used to be the most challenging part of creative work. Creators put in lots of thought at the beginning to explore and generate ideas. Once an idea is settled, the execution process starts, which can be slow and tedious. There are many details to hammer out, but progress is controllable and steady.

As demonstrated by the old 80/20 pattern, achieving the first 80 percent in relation to perfection required the most work:

figuring out ideas and concepts, and producing the rough first version. The remaining 20 percent involved refinement: polishing prose, fixing edge cases, tightening the design. That final stretch was time-consuming and often frustrating, but the creator knew what needed to be done. The path to the final output was known. To use a hiking metaphor, the difficulty was in enduring the long walk, not in finding the trail.

With AI, huge progress can occur instantly. A single prompt generates a draft, a sketch, or a block of code in seconds, compressing hours of work into minutes. That is why the old 80/20 logic no longer fits. The new pattern is closer to 90/10. However, the deeper change is not the ratio; it's the nature of the production process. AI hands you something that looks nearly finished before you have worked out what is wrong with it, before you know what "better" even means, and before you know which of the many possible improvements would matter. The creative part of the production now comes later in the process.

Instead of laboring on a long hike, AI works like a helicopter. AI drops you near the summit without the long climb. You save enormous effort, but you also skip the accumulation of skill and judgment. You miss out on the feel that the climb once provided, and you do not see the trail that leads to the summit. Nevertheless, you can still ask AI for help for the last part of the journey. This creates two opposing forces.

The knowledge deficit. In the traditional production process, getting to 80 percent of perfection was evidence of the worker's knowledge. A writer who had produced a strong draft had already wrestled with the structure of the argument and was knowledgeable about the supporting materials. As she began her revisions, she knew all the foundational materials and only needed to fix the surface.

With AI, a draft appears before the thinking is done. The writer now faces decent prose without an equally developed mental model behind it. Her mind is no longer ahead of what's written on the page, but behind it. So the hardest part of the job, including diagnosing weaknesses, judging coherence, and deciding what matters, is harder than before. That is why AI so often produces stagnation: The user falls into a cycle of prompt, glance, and reprompt. New output arrives, but the work does not improve.

Who breaks through the stagnation depends on two factors. One is existing skill and domain knowledge. A skillful editor knows the structural reasons why a draft is wrong. A less experienced one may not. An understanding of what excellence looks like is a key factor for breakthrough. The second factor has to do with the willingness to do hard thinking. The availability of AI makes it easy to outsource the thinking prematurely. And real improvement still requires hard thinking.

The companion effect. AI may create the knowledge deficit that impedes further progress, but it also gives something back. In the past, a writer stuck at 90 percent was often alone with her doubts. Writing alone can be mentally taxing. Now, with AI, the writer can ask the program to critique her argument, suggest alternative framings, identify the weakest points, and generate different ways to restructure the conclusion.

This means the marginal cost of continuing to work has fallen. When AI functions as a thinking partner, the next hour of effort is less unpleasant and less lonely. Writers may even develop a "flow experience" while working with AI, continuously pushing forward the quality of the writing.

In summary, AI raises the difficulty of the last 10 percent because the creator lacks the accumulated understanding that she would have gained if she had endured the long, hard climb. AI simultaneously lowers the cost of working on the last 10 percent because there is an algorithmic collaborator. That combination creates a bifurcation: Some people stop working as soon as the initial gains of AI end, but others keep working because they know they can improve, and because the marginal cost of working more is acceptable. The evidence shows both patterns.

The mediocrity trap and the last-mile illusion

To examine the effect of AI on the creative process, Jin and his fellow researchers ran an experiment with 219 artists, ranging from students to professionals. The task was to illustrate passages from famous novels, both with AI and without it. Each artist worked for up to 150 minutes. The researchers looked at their entire creative process.[178] They recorded every brushstroke, undo, and prompt. Independent judges scored the artwork at fifteen-minute intervals. Without AI, the pattern was the familiar one: gradual progress with diminishing returns. The first hour produced a rough composition. The artists then added color and shading, and finally worked on fine details.

With AI, the shape of progress changed sharply. Within thirty minutes, many artists reached a quality level that would have taken two hours by hand. Then they hit a wall. After forty-five minutes, the quality curve started to plateau. Further gains were barely noticeable, even as artists kept tweaking prompts and patching details. This pattern is consistent with the knowledge deficit discussed earlier. AI helped the artists to get a polished image before they had thought enough about the composition, which made it difficult to make improvements.

The new shape of progress led to behavioral changes in many artists. Without AI, almost all artists worked to the deadline. With AI, there was a bifurcation. One group stopped early, right around minute sixty, where the curve fully plateaus. They reached a point where additional effort no longer seemed to improve the work, so they decided to stop. For them, the expected marginal benefit had fallen sharply. Access to AI did not bring in higher final quality ratings for these artists. AI became a time-saver but it also created a mediocrity trap that stopped them.

The other group did the opposite. They kept working all the way to the full 150 minutes. This extra time, however, did not significantly improve the arts' quality. The final quality ratings were not noticeably different from those who stopped earlier. In other words, the quality plateaued in the time allowed. This was the group with low marginal costs.

In art, "good enough" is sometimes acceptable. But what happens in settings where precision is more important, such as software engineering? Model Evaluation & Threat Research (METR) explored this question in a software development field experiment. They recruited sixteen experienced open-source developers and gave them real tasks drawn from large, mature codebases that the developers already knew well. These were projects for which the contributors had, on average, years of prior experience. Tasks were randomly assigned to an "AI allowed" condition or a "no AI" condition. The researchers measured how long the same kinds of tasks took to complete in each mode.[179]

The result was the opposite of what almost everyone expected. With access to frontier coding assistants, developers took about 19 percent longer to finish their tasks. Interestingly, the developers *felt* like they completed the work faster. Before starting, they predicted a substantial speedup, and afterward they still believed that AI had helped them. But the stopwatch showed the reverse.

This finding is in line with the two effects in the "helicopter drop" logic. AI gets you to the high-altitude camp instantly, generating 90 percent of the code quickly. But because you did not make the climb yourself, you lack the knowledge that would let you finish the last mile quickly. At the same time, the companion effect is still present. AI lowers the cost of continuing to work, and so you keep going because bugs need to be fixed. Because the rate of progress is lower, and because the programmers won't stop until they reach a sufficiently high level of output, the entire process takes even longer than before.

The organizational standard-setting problem

The 90/10 production function creates a problem that is both organizational and individual. When AI gets everyone to 90 percent instantly, the organization faces a new sorting challenge: How do you distinguish workers who pushed to 98 percent from those who stopped at 91 percent? The surface looks identical. The verification costs we analyzed in chapter 10 make the problem worse because the manager who reviews the output cannot easily tell which version represents serious effort.

The complementarity logic from chapter 7 applies here. An organization's quality standards, incentive systems, and evaluation processes form a cluster. If the production function changes trajectory from a slow climb to a helicopter drop with a plateau, the entire cluster must adjust. An organization that still rewards volumes of polished output is rewarding the 90 percent that AI provides for free.

The first step is to stop treating all output equally. An organization's work can be roughly divided into two categories of tasks. The first is *commodity tasks,* such as preparing meeting minutes, formatting data for spreadsheets, or writing routine status updates. In these cases, a 90 percent quality rating is good enough. The second category is *star tasks,* such as a client presentation or a high-stakes negotiation, where success requires the last 10 percent as well.

For commodity tasks, the goal is competence, not excellence. For these tasks, if the AI gets you to 90 percent, the organization should accept that and move on. Spending an hour to polish it further is a waste. Star tasks are quite different. For these, the difference between good and excellent is crucial. The organization needs to invest the time to elevate the work from acceptable to excellent.

The harder step is to redesign incentives so that workers who complete the last 10 percent on star tasks with excellence are rewarded. The company should not reward volume across all tasks. An organization that only measures productivity by volume will get mountains of 90 percent work and very little excellence.

For the individual, the implications are the same. You should sort your own work into commodity and star tasks before you start.

The mistake many people make is treating every task like a star task.

This brings us back to where we began. Jason Allen did not win at the Colorado State Fair by pressing a button. He won because he treated the piece as a star task and refused to stop at the plateau. His eighty hours of prompt iteration and twenty hours of manual refinement were not wasted effort; they were the price of the last 10 percent. Most people who generate an image with AI look at it, feel impressed, and move on. Allen looked at it, saw what was missing, and kept going. For organizations, the question is whether their systems can identify and reward the Allen in the room, or whether they instead treat his hundred hours of work the same as someone else's five minutes.

CHAPTER 12

The Career Ladder Is Broken

How AI Closed the Training Pipeline and How to Rebuild It

In *The Skill Code*, organizational scholar Matthew Beane describes what happened when robotic surgery systems entered hospital operating rooms. The technology was extraordinary for senior surgeons: precise, efficient, and "iPhone-easy" to use. But it was a disaster for training. Robotic systems concentrate the procedure in the hands of a single operator at a console. The attending surgeon has no reason to share control with a trainee because doing so means a slower operation and greater risk to the patient. So residents stand to the side and watch. But watching a movie does not make one an actor. The residents never developed the hands-on skills that surgeons always built through years of practice.[180]

The consequences show up when the residents become independent surgeons. In one case, a young surgeon named Kristen sweated at the console when she started her first independent job. She hesitated and, as a result, burned and cut a lot of extra tissue. The surgery, which would have taken three hours, took seven. The patient lost about ten times more blood than otherwise would have occurred.

This problem is not limited to the medical sector. In 2024, a tech executive told *New York Times* columnist Kevin Roose that his company had stopped hiring software engineers below L5, which is a mid-level title typically requiring three-to-seven years of experience. The reason was simple: AI coding tools could handle lower-level tasks. Another founder reported that his startup employed a single data scientist to do work that required a team of seventy-five at his previous company.[181] *The Wall Street Journal* reported that marketing agencies had all but stopped requesting entry-level staff.[182] Kate Smaje, McKinsey's head of technology and AI, put it bluntly to Bloomberg in 2025: "Do we need armies of business analysts creating PowerPoints? No, technology could do that."[183]

These are not isolated anecdotes, but they overlook the need to train younger workers. Although the evidence is still emerging, SignalFire found that the hiring of new graduates had fallen 25 percent.[184] The New York Fed documented youth unemployment rising 30 percent since 2020, twice the rate of all workers.[185] This happens because the economic logic that sustained those positions has changed. AI can now do some work that junior employees used to do, the work that taught them to become senior employees.

Economists Lichtinger and Hosseini Maasoum gave this pattern a name: seniority-biased technological change. Analyzing millions of resumes, they found that productivity gains accrue to those who already possess expertise while junior hiring drops. The technology helps experts but eliminates many positions through which people become experts.[186]

The importance of learning by doing

To understand what AI is breaking, we must first look at how professional expertise is actually forged in the real world. Formal education provides the theoretical foundation of a profession, but mastery requires physical execution. When we look at the data across entirely different professions, a clear empirical reality emerges: Expertise is the direct byproduct of doing the work.

In a landmark 1962 paper, the Nobel Prize-winning economist Kenneth Arrow provided a fascinating review of empirical evidence to demonstrate this.[187] He pointed to the Horndal ironworks in Sweden, a plant that had seen no new investment in machinery or changes in production methods for fifteen years. Yet, over that same period, output per worker steadily rose by 2 percent annually. The workers and managers simply got better at the process through repetition. Arrow also highlighted a famous 1936 study of airplane airframes, which showed a remarkable and highly predictable phenomenon: The number of labor hours required to assemble an airframe declined steadily as the total number of airframes previously produced increased. Engineers and assembly line workers weren't taking breaks to study aeronautics textbooks; they were getting faster and better simply by building more planes. The physical act of production was the mechanism of learning.

A few years after Arrow's paper, Bruce Henderson and the Boston Consulting Group (BCG) popularized this phenomenon into the famous "experience curve," demonstrating that across many industries, a company's unit costs predictably fall by a fixed percentage every time its cumulative production volume doubles.

Entire organizations learn by doing.[188] Decades later, Malcolm Gladwell coined the individual equivalent of this concept with his (not scientific, but catchy) "10,000-hour rule." Whether you are an aviation plant, a global consultancy, or a single human being, you cannot read your way to expertise; you have to put in the work.[189]

The importance of on-the-job training also holds true in professional sports. Researchers examined the career trajectories of English soccer players whose teams were narrowly relegated to a lower division.[190] Intuitively, dropping to a lesser league seems like a devastating career setback. However, because relegated teams face severe budget cuts, they can no longer afford expensive veteran stars. Out of necessity, they thrust their younger, less-experienced players onto the field. Because these rookies got to play rather than warm the bench for an elite team, they accumulated so much human capital that, five-to-seven years later, they ended up playing in better leagues and earning higher wages. Doing the work built the career.

Conversely, when professionals are denied the opportunity to practice, their skills decay. Researchers looked at public school teachers in Greece who were placed on long waitlists before receiving formal teaching assignments. They calculated a skill depreciation rate of 4.3 percent per year.[191] When human capital is not used, it atrophies. Missing out on the early returns of on-the-job experience leaves a permanent scar on a professional's capability.

Real experience is essential, but young workers need to be guided by expert mentorship to develop true mastery. When a novice works alongside a master, tacit knowledge, such as tastes and problem-solving heuristics, is transferred. A fascinating study on Western music composition since 1450 proves just how durable this imprint is. By analyzing the musical structures of thousands of compositions over five centuries, the study found that students absorb and replicate the distinct stylistic elements of their direct mentors. Furthermore, students who emulate high-quality mentors are significantly more likely to become top-tier composers themselves, passing down this "acquired taste" across generations.[192]

The examples above illustrate the importance of real-world experience and training. But if expertise requires a massive amount of practice guided by experts, how do firms afford to let novices work enough to learn without wasting the precious time of the experts?

Knowledge hierarchy: two birds with one stone

The answer lies in how work itself is organized. In any profession, tasks range from the routine to exceptional, from reviewing standard contracts to restructuring a bankrupt conglomerate. This diversity in difficulty is the key. If every task required senior expertise, there would be no way to let novices practice without the guidance of experts. But most tasks do not require high-level expertise. Most problems at a law firm, a hospital, or a consulting practice are common, well-understood, and low stakes. A much smaller number of problems are unusual.

This natural distribution of difficulty makes it possible to build an organization that kills two birds with one stone. The structure is the knowledge hierarchy, which we touched on earlier in the book. The knowledge hierarchy enables the organization to use expertise efficiently and train the next generation at the same time.

To use expertise efficiently, organizations should not require the most expensive expert to answer the easiest questions. It is better to push routine work down and funnel exceptions up. Juniors handle the common, low-stakes problems at the bottom. When they cannot solve something, they escalate. Rare, high-stakes exceptions flow to fewer, more knowledgeable experts at the top. When partners can focus their attention on the hardest problems, the knowledge hierarchy leverages their expertise.

The knowledge hierarchy serves a second, equally important function: It trains people. It would seem natural for firms to simply train their workers, but economist Gary Becker identified a fundamental problem with workplace training.[193] It creates valuable skills, and valuable skills command higher wages on the open market. This is particularly the case when the skills being transferred are generally applicable across firms. If a firm invests time and money into training a worker today, competition will force the firm to pay a higher wage tomorrow to retain her. Unless the firm has some form of monopsony power (the ability to restrict workers' mobility and pay them less than their market value), it will be unwilling to train them. Otherwise, rivals can simply swoop in, offer a higher wage, and poach the newly trained employee, reaping the benefits without paying for her education.

To solve this, the junior workers would need to pay for the training, since they are the ones who will reap the future wage benefits. But young workers are financially constrained. They have rent, groceries, and student loans to pay. Firms can hardly charge a junior employee a "negative wage" to cover the cost of the time senior experts spend mentoring them.

The knowledge hierarchy solves this problem through an elegant bargain that one of us (Garicano) developed in 2017 with Northwestern University economist Luis Rayo. Unable to pay for their apprenticeship with cash, junior workers pay with their labor. Because they are not yet highly productive, the arrangement has them perform massive volumes of routine, foundational work at the bottom of the pyramid. They accept lower wages than their output justifies, generating revenue for the firm while they learn.

Consider again that first-year associate. She bills at $400 per hour for document reviews. The client pays for productive work, not training, but this revenue exceeds her salary and costs. The surplus subsidizes the time that partners spend mentoring her, the training programs she attends, and the learning that occurs as she tackles progressively harder work.

The associate accepts lower wages than her output justifies in exchange for skills that will command higher wages later. The firm extracts immediate value from her work, recouping training costs before she might leave for a competitor. Both sides benefit because the same activity produces two outputs: The associate reviewing contracts is doing billable work and learning to spot problems. The work pays for the learning while the learning makes future work more valuable.

This bargain has held since medieval apprenticeships. The apprentice worked for room and board while learning the craft. The master captured cheap labor while transferring skills. The arrangement worked because learning and productive work were inseparable.

How AI changes the career ladder

The AI-Becker problem

From the standpoint of productivity, delegating routine work to AI looks like a clear improvement. The machine absorbs low-level questions and ensures that scarce expert attention is reserved for the hardest problems. McKinsey's AI tool Lilli now answers half a million queries a month, generating the initial research on a project that once occupied hundreds of junior consultants. Allen & Overy's AI tools automate contract drafting and due diligence that first-year associates used to perform during thousands of hours.

But this efficiency gain comes with a hidden cost. If a machine can do the routine work cheaply, junior employees lose the "currency" they used to pay for their training. In other words, they no longer have anything of immediate economic value to offer the firm in exchange for their training.

We call this the AI-Becker problem.[194] In Gary Becker's original analysis, companies may underinvest in training because rivals can poach trained workers. In the AI era, companies underinvest because juniors simply cannot generate the revenue required to fund their own apprenticeships. As Garicano found in recent research with Luis Rayo, the currency an apprentice previously offered for her training (the menial, routine work) depreciates when AI can perform these tasks.[195]

Beane documented this pattern in robotic surgery, which we described at the opening of this chapter. The technology that made the expert more productive blocked the path through which new experts were made. The same dynamic is now playing out across knowledge work. If AI drafts contracts, builds financial models, and generates the research that junior consultants used to produce, those juniors will never get the chance to do the work themselves. And without opportunities to struggle through a problem, make mistakes, and learn to recognize what went wrong, they cannot develop the expertise and judgment that the work was supposed to build.

The delayed catastrophe

Forward-looking executives are already sounding the alarm about the consequences of hollowing out this pipeline. AWS CEO Matt Garman recently criticized the short-sighted impulse to cut entry-level roles to capture immediate savings. "How's that going to work when ten years in the future you have no one that has learned anything?"[196]

The crisis will not arrive immediately, which is both reassuring and dangerous. Today's senior professionals were trained traditionally. They possess judgment that AI lacks. They have the ability to recognize when AI errs, to spot the exception that requires human attention, to direct AI toward the right problems. With AI assistance, they are more productive than ever. Because this expertise has a long shelf life, firms can function for years without noticing that they are no longer investing in a future generation of experts. The senior workers will retire eventually, and when that occurs in five or ten years, firms will realize the pipeline of mid-level talent is empty. We are consuming expertise without replanting it.

What determines whether careers survive is a race between two forces. AI raises the floor by automating entry-level work. But AI also raises the ceiling by making expert work more valuable. If the ceiling rises faster than the floor, there will continue to be space for training and the career will still be worth pursuing. If the floor rises faster than the ceiling, then training will compress and may eventually collapse. Current evidence suggests the floor may be rising faster than the ceiling. Entry-level hiring is freezing while senior hiring and retention remains stable. Productivity gains are accruing mostly to those who already possess expertise.

How organizations can rebuild their pipelines

Organizations must either invest in developing new talent or face the delayed catastrophe that Matt Garman described. Three responses are possible, though none is easy.

First, organizations can restructure junior roles around verification and quality assurance rather than routine production. If AI produces the first draft, the junior worker's job shifts from drafting to checking citations and assumptions, and catching potential errors. This work is difficult, but it builds real expertise and it is hard to automate. It also helps to solve a second problem, which is the verifier's curse. It does this by putting fresh eyes on AI-generated output. The junior learns by checking and the organization gets the quality assurance it needs. The economic bargain is preserved because the junior worker's verification efforts pay for her training.

Second, organizations can invest in rotations and shadowing

programs that expose juniors to the messy, political, relational work described in chapters 6 and 7. If the routine cognitive tasks that once served as training currency disappear, then the training must shift to the tasks that remain, such as coordination and implementation. These skills are acquired only through practice, but they can be practiced alongside experienced professionals if the organization deliberately creates those opportunities.

Third, most organizations do not know what AI can do, which recent college graduates do know because they are AI native. Organizations could allow junior employees to make proposals and offer to work on automating exiting workflows through the use of AI.

For the individual worker, the organizational logic determines the strategy. If you are early in your career or trying to start a second one, AI will be a serious challenge. On the traditional path, you would have expected to grind through routine work for several years, use your labor as "currency" to pay for the training, and eventually graduate to more senior responsibilities. That path is closing. You can no longer wait for training to come to you. Some alternatives are provided by AI itself.

One example is to use AI as a tutor. Anyone with an internet connection has access to a tutor of extraordinary patience and enormous knowledge that will explain a concept in ten different ways (without irritation), adjust to any level, and work across every domain. Young professionals can use the system to build deep knowledge and learn to learn. As models get better, fewer humans will be good enough to add value at the routine level. To exercise judgment, you must know a domain deeply, but the specific knowledge you learn today will depreciate faster than ever. What matters is the slope of your learning curve. Choose the job where you will learn the most, but also the one where you will learn to learn.

This does not solve the AI-Becker problem. A tutor is not a career. It cannot give you the clients, the deadlines, the colleagues, or the network. But it allows you to pursue ambitious strategies that would have been unthinkable a little while ago. Here are some ideas.

Entrepreneurship. There has perhaps never been a better time to strike out on your own. In the past, becoming an entrepreneur early in a career was daunting because it required you to be a jack of all trades. You needed to code, design, sell, keep books, and understand your market, all before you had revenue to hire anyone. The jack-of-all-trades problem killed most ventures before they found product-market fit. You often failed at whichever was your weakest link.

Today, as we discussed in chapter 4, AI relaxes this constraint. You can use AI to handle your weak links while you focus on core tasks. If established firms will no longer invest in your training, you can train yourself by building something of your own.

The smaller pond. Historically, landing an entry-level job at a prestigious bank, hospital, or consultancy almost guaranteed a good career because the on-the-job learning was best. But if elite firms are deploying AI for entry-level tasks and sidelining junior workers to merely watch the machine work, that premium vanishes. Watching a senior doctor operate on a patient at a console does not make you an expert.

A smaller pond may be a good alternative route. This might be a regional firm or a niche where you are slightly overqualified and therefore trusted with real responsibility from the start. In those environments, you will get to do things and learn from practical experience. Remember the lesson from English soccer players. They succeeded in the long run because dropping to a lower league gave them more experience in the game.

Seek scarcity. A top corporate lawyer at the peak of his discipline is rare but not unique. A top corporate lawyer who also understands machine learning well enough to structure deals around AI assets is almost nonexistent. Many people have specific skills but almost no one has a combination of AI technology knowledge and domain-specific capacity. Scarcity lives at those intersections. Scarcity does not live at the peak of a single discipline. These "scarcity intersections" are where the messy jobs can be found. They require the ability to manage unsolved problems, exceptions, and conflicts. A person who can operate credibly on both sides of the AI threshold will be harder to replace. In the past, building depth in two domains meant two long apprenticeships. Now the AI tutor reduces those learning curves.

CONCLUSION

The Economy of Messy Jobs

Up to now, we have focused on AI's impact on individuals, organizations, and their messy jobs. Our discussion also raises questions about the economy as a whole: Will AI create mass prosperity or mass unemployment? Will AI move more transactions to the markets or to firms? Will AI produce explosive economic growth or stagnation?

Media headlines love to scream the answer given by some of the smartest people in tech, which is that we are approaching an economic "singularity." They believe that AI will invent better AI, which will invent still better AI, producing exponential improvements in living standards. Within a few years, they say, scarcity will end and humanity will enter an era of abundance.[197]

This extreme optimism is unrealistic and disturbing. Its proponents see the economy as a machine built and run by advanced technology. They ignore the fact that the economy is a web of messy jobs, derived from human needs, organizational frictions, institutional rules, and social expectations that no technology, however powerful, can optimize from the outside.

Consider what we have seen already. The radiologist's job survives automation because his real work involves persuading surgeons and comforting patients (chapter 3). The entrepreneur discovers new problems that are now feasible and financially worthwhile to solve (chapter 4). The nurse practitioner's role expands because AI breaks through her cognitive ceiling while leaving the messy relational care untouched (chapter 5). The manager spends her days navigating organizational politics and grappling with decision ambiguity (chapter 6). The implementer bundles processes, technology, and political knowledge to manage architectural change (chapter 7). The master engraver's work commands a premium because the human origin itself cannot be replicated by a machine (chapter 8).

The messiness of real jobs is part of human nature and part of institutional evolution, as we have argued throughout this book. It resists the blind adoption of technology, slows the macroeconomic transition, allows society to resolve distributional consequences, and creates space to develop regulations that protect human value and avoid catastrophic outcomes.

We also disagree with pessimists who see AI as the latest in a long series of overhyped technologies that generate excitement in Silicon Valley but disappointment in governments' national coffers.[198] This camp ignores the ways in which AI differs from every previous technology. AI is an industrial revolution of *cognition.* It can solve a wide range of challenging cognitive problems. It can raise the floor for every type of knowledge worker rather than benefit just a subset of skilled workers. It enables talented individuals to leverage scarce resources and it empowers newcomers to materialize their new ideas.

Our view is more hopeful than either camp. It is grounded in two concepts: the messiness of jobs (the irreducible tangle of judgment, context, relationships, and improvisation that constitutes most real work), and the autonomy threshold (the line beyond which a machine can perform a task without meaningful human oversight). Together they determine which tasks machines take over, which remain human, and what happens to growth, wages, and opportunity when a powerful general-purpose technology meets the real world.

This book's central argument points to a more optimistic future than either of the extreme views. Its aggregate implications will play out through three channels. The first is the *production of goods*: how AI reshapes work across the three zones of clean automated tasks, messy bundled tasks, and irreducibly human tasks. The second is the *boundary of the firm*: which transactions move toward markets and which stay inside organizations. The third is the *production of ideas,* which determines whether AI lifts the long-run rate of growth.

Prosperity through more efficient production and reallocation

The first channel through which AI affects economic growth is the production of goods and services. This channel operates through automation and reallocation. AI substitutes for human labor in some tasks while redirecting economic activity toward others. To understand the overall effect, we examine each zone of the autonomy threshold separately.

Tier one: the automation zone

AI substitutes for human labor in tasks that sit above the autonomy threshold, as we discussed in chapter 2. When a firm deploys AI to automate tasks, the firm can do more work with fewer people. Economists call this *capital deepening*. As a result, output per worker rises, wages tend to go up, and costs and prices fall.

Capital deepening by itself cannot sustain permanent growth. Each additional machine produces a bit less than the last. This is the central lesson of the neoclassical (Solow) growth model: capital accumulation alone produces a one-time increase in the level of output, not a permanent increase in the rate of growth.

Tier two: the strong bundles

The second zone contains the vast middle of the economy, the messy normal jobs where AI automates only some tasks within the strong bundle and operates below the autonomy threshold at the job level, as we discuss in chapter 3. The nurse uses AI diagnostics but still holds the patient's hand. The restaurant manager uses AI scheduling but still leads a kitchen crew. These workers will not be displaced. They will be augmented.

The process of task unbundling and rebundling due to AI will improve the efficiency of the economy. The "robot below" model of chapter 4 enhances the leverage of top talent while the "robots above" or "expertise in a box" model of chapter 5 allows average human workers to use the exceptional knowledge encoded by AI.

By helping to patch the weak links of the economy, AI can create substantial development dividends. The scarcity of expertise is the binding constraint on development. A poor country has few doctors, few professors, and few engineers. With AI tools, expert knowledge becomes accessible to people who could not previously access it. For instance, AI diagnostic tools allow a nurse in rural

Kenya to handle cases that previously required a specialist in Nairobi. Through AI, farmers in India may have access to a top agronomist's know-how that can radically improve their productivity.

Tier three: the human-dominant zone

The third zone contains the messiest jobs, work that is deeply entangled with human judgment, creativity, persuasion, and accountability. Think about political negotiation, management, and implementation (chapters 6 and 7); or craft, performance, care, and service whose value depends on human origin (chapter 8). These jobs require more than cognitive skills. They also require character, stubbornness, courage, and the willingness to be held personally responsible for unpredictable outcomes. AI cannot supply any of this.

Crucially, the demand for human work in this third zone is likely to grow as the other two zones become more productive. A society with abundant cognition does not stop spending money. It redirects spending toward services with human presence. These services include healthcare with more attentive nurses, restaurants with more staff per table, schools with smaller classes, and cultural experiences with more human performers. The entire range of relational and embodied work we cataloged in part II will see its economic weight rise. The Baumol disease becomes a Baumol blessing.

This zone may also serve as a brake on the displacement of human jobs. The surgeon cannot operate faster just because a better AI identified the tumor. The therapist won't be able to compress a session even if an algorithm wrote the treatment plan. As prices in automated sectors fall, the share of spending in human-intensive sectors will grow. Even though the slow productivity growth of human-intensive sectors will pull the economy-wide average down, that slower pace will help protect workers from losing their working hours to machines.

Shared prosperity after a painful transition

Having aggregated the responses to AI from the three zones, we foresee a future that is very different from the IT revolution. Information technology rewarded scarce analytical skills and hollowed out routine cognitive work. AI is not an inherently egalitarian technology. But AI, when deployed below the autonomy threshold as a copilot, gives everyone access to cognitive tools previously reserved for the highly trained. Moreover, some of the scarcities created by AI, such as physical presence or relational skills, may be more broadly distributed than the analytical skills made valuable by the IT revolution. If this pattern holds, the distributional signature may be compression rather than polarization.

Some jobs will be directly displaced by AI. The workers who hold them will need to move, just as agricultural laborers moved into manufacturing and manufacturing workers moved into services. The transition will be painful for the generation that lives through it, and the pain must be taken seriously. But history tells us that the economy creates new work to absorb the displaced, *if* institutions support the transition.

The changing role of firms in the AI era

The messy jobs logic also helps us think about the boundary of the firm; that is, which transactions are organized inside firms and which are left to markets. Ronald Coase asked why firms exist at all when prices can coordinate economic activity. His answer was that using the market is costly. It involves finding counterparties, writing contracts, monitoring performance, and resolving disputes. AI lowers many of these costs. Humans can use agents to search for suppliers, compare bids, draft agreements, connect software systems, and execute routine transactions. This has led many in Silicon Valley to predict what they term the Coase Singularity. They believe that if AI agents reduce sharply the cost of using markets, the boundary of the firm will retreat. Activities that were brought inside firms because market coordination was too costly should move back outside. For instance, Stripe cofounder John Collison has predicted that there should be "fewer people per firm, more output per firm, more firms, and more coordination through market-like mechanisms."[199]

Our analysis has clear implications for when markets will advance and when firm boundaries will retreat, and when the opposite will happen. Chapter 4 showed the first half of the answer, consistent with the Silicon Valley view: When AI solves the routine problems, the entrepreneur can accomplish much more without hiring specialists. The highly skilled expert can leverage her skills through thousands of agents. A small team can now assemble a bundle of tasks involving production, sales, support, finance, design, and legal work that once required multiple departments. This should mean more entry, more niches, and more one-person or small-team firms.

The other half of the answer is harder. Many transaction costs are not related to search, or to executing and creating processes. Instead, the costs are related to strategic disclosure, commitment, adaptation, authority, and trust. These are political and institutional costs. They show up in the cases we examined in chapter 6. When xAI needed an emergency grid connection in Memphis, no agent could force the local utility to reveal its true costs because the utility had every reason to inflate them. Economic theory shows that no negotiation mechanism can simultaneously deliver honest reporting, efficient allocation, and voluntary participation when parties hold private information and have conflicting interests. AI does not change this reality. The constraint is strategic, not cognitive. An agent can make it cheaper to draft contracts, but it cannot force an unwilling counterparty to reveal private information or trust a process.

These costs are central to what we described in chapters 9 to 12. Those chapters show that firms still play a critical role as repositories of trust and nexuses of long-term relationships.[200] In chapter 9 we described how AI compresses cheap signals, which means that organizations must rely more on costly, verifiable histories. Firms create durable relationships in which reputations are built and sanctioned. Next, in chapter 10 we showed how fluent output creates workslop, which raises the value of teams whose members trust each other to do real work. Long-lasting reputational interactions between workers, in the form of relational contracts, are something firms do well. In chapter 11, we demonstrated how organizations assume responsibility for production standards and determine what "good enough" means. And in chapter 12 we argued that firms can serve as training systems by developing long-term relational contracts between apprentices and masters that transfer knowledge in exchange for work.

AI cheapens what spot markets already do well: search, contracting, execution, and the verification of standardized output. Markets do not make trust, reputation, authority, or accountability instantly tradable. The institutions that host relational activities will gain ground. And firms, not markets, remain the most common home for messy, bundled, long-term work.

Hence, the likely result of these changes is not the disappearance of the firm but a change in its job description. Clean, modular, verifiable tasks will move toward markets and agent-to-agent contracting. Messy bundles will stay inside firms or durable teams because those bundles require authority and trust. The firm of the AI era will employ fewer people to process routine cognition, but it will matter more as the place where reputations are built and accountability is borne.

More ideas but not explosive growth

AI will improve the efficiency of the economy through automation and task reallocation. This will create an enormous level gain, probably in the same way that the mechanization of agriculture did. But the level effect will be limited. We will settle in a better place, but we will not see continuing faster growth as a result.

Whether the economy can permanently grow faster depends on whether AI helps us generate more and better ideas. Economists often disagree with each other, but they share one piece of common wisdom: Ideas are the ultimate source of long-run growth.[201]

History shows again and again that physical constraints (limited energy, space, food) are overcome only when humans come up with new ideas. AI has the potential to transform the process of idea generation and R&D activities.

Many R&D activities start by finding new combinations of existing tasks, and then experimenting with them. AI plays a powerful role in this process if the rules for combinations are well-defined. After retrieving millions of studies, it can suggest cross-disciplinary combinations that human experts cannot match and quickly test the most effective combinations. This is the success of AlphaFold, a model developed by DeepMind that won the 2024 Nobel Prize in chemistry for solving the protein folding problem.

While AI can automate some simple research tasks, scientific discovery is largely a messy process. Discovering a candidate molecule is different from making a new drug. It must be synthesized in a physical laboratory and tested for toxicity. It must undergo three phases of clinical trials. Each step requires a series of tasks that involve significant human intervention, such as finding patients, obtaining informed consent, monitoring outcomes, managing adverse effects, and communicating with regulatory agencies. This tedious pattern is repeated in all areas of innovation. AI accelerates the generation of ideas, but evaluating and implementing these ideas remains a human endeavor.

The most challenging part of knowledge creation is not finding answers, but deciding which questions are worth asking and then mobilizing resources to investigate them. Before deciding whether to develop a drug, pharmaceutical company executives must understand patient populations, the political climate, and even consider potential bias from FDA reviewers who ultimately make the decision. Cheaper cognition does not necessarily make this process easier.[202]

Generating new ideas and translating them into actual products requires a series of interdependent innovative tasks. Until AI excels at every task, the problem of weak links will persist. For example, AI can search the space of protein structures and candidate molecules at speeds that human teams cannot match. However, even a thousandfold increase in search speed cannot speed up clinical trials, which determines how quickly new drugs can reach patients.

Weak links will have an impact on every sector: materials, energy, education, construction, and public health. In every case, the automated step becomes cheaper, and the human-dominant steps bind tighter as everything around them moves faster. Thus, a world with cheaper cognition will demand more irreducibly human work, such as resolving conflicts, negotiating between stakeholders, redistributing benefits and costs, and building trusted public institutions. This type of messy human work has been scarce, and it will be even more scarce when the cognitive tasks around them become nearly free and the queue of ideas waiting for human judgment grows.

Three forces, each rooted in the autonomy threshold logic of this book, will produce fast but bounded growth. First, the autonomy threshold for ideas means that discovering which problems matter will remain human work. Second, the cost of implementation means that deploying any new idea will require messy organizational change that cannot be automated. Third, the economic weight of human work has increased. As a result, even when automated sectors surge, the Baumol effect will pull the economy-wide average down.

The result is a rate of economic growth that is neither explosive nor stagnant. It will be faster than the last three decades, but not exponential. On one hand, AI partially relieves the burden of knowledge. On the other hand, the messy human work of choosing problems, implementing solutions, and delivering human-intensive services cannot be accelerated by computation alone.

We have no doubt that an economy where cognition is cheaper will produce more ideas and grow faster. Nevertheless, messiness will protect human jobs because the economy requires some features AI cannot supply. Messiness shapes the economic growth path because it allocates the gains to human-intensive sectors and keeps the innovation process dependent on human choices. This will lead to an economy that extends its benefits more broadly across rich and poor countries.

The world is not running out of problems that need solutions. Disease, poverty, climate change, and an aging population are not problems that language models can solve on their own. They require human action, human determination, human accountability, and human stubbornness. If you bring those qualities to work that matters, you will not be replaced. You will be needed more than ever.

Acknowledgements

Thanks to Daron Acemoglu, David Autor, Matthias Fahn, Zhiqi Feng, Pieter Garicano, Jie Gong, Evan Guo, Jienan Han, Jiayi Hou, Yao Huang, Alex Imas, Hanzhe Li, Raffaella Sadun, Eduard Talamàs, Jesús Saa-Requejo, Vicky Wan, Kunyi Wang, Richard Wong, Zhenyu Zhu, and Andrew Zhang. Very special thanks to our wonderful editor Glenn McMahan.

About the Authors

Luis Garicano is a professor of public policy at the London School of Economics. His research studies the organization of knowledge, talent, and specialization, and how computerization reshapes firms, productivity, and inequality. His recent work has focused on the impact of AI on careers, jobs, and on how firms can succeed at AI implementation. He cowrites the Substack newsletter *Silicon Continent*, which is about AI, innovation, and European technology policy. From 2019 to 2022, he served as a member of the European Parliament, where he led economic affairs for the Renew Europe group. He has a PhD in economics from the University of Chicago where he was a professor for a decade. He has also taught at MIT Sloan School of Management, Columbia University Graduate School of Business, and London Business School.

Jin Li is the Zhang Yonghong Professor in Economics and Strategy, director of the Centre for AI, Management and Organisation, and Area Head of Management and Strategy at the University of Hong Kong. He is an editor at *The Economic Journal* and an associate editor at *Management Science*. His research spans organizational economics and strategy. His recent work focuses on AI adoption, AI and organizational transformation, and incentive issues arising from AI. He received his PhD in economics from MIT and has taught at the London School of Economics and Northwestern University.

Yanhui Wu is a professor of economics, and of management and strategy, at the University of Hong Kong, where he serves as Area

Head of Economics and directs the Digital Economy Research Institute (Shenzhen) at the HKU Business School. His research spans organizational economics, media economics, and the digital economy. His recent work focuses on how AI is reshaping jobs, education, and well-being. He received his PhD in economics from the London School of Economics.

Endnotes

1 Lloyd Lee and Kelsey Vlamis, "Microsoft AI CEO Predicts 'Most, If Not All' White-Collar Tasks Will Be Automated by AI within 18 Months," Business Insider, February 12, 2026, https://www.businessinsider.com/microsoft-ai-ceo-mustafa-suleyman-white-collar-tasks-automation-prediction-2026-2.

2 Jim VandeHei and Mike Allen report that Dario Amodei said that AI could wipe out half of all entry-level white-collar jobs and spike unemployment to 10 to 20 percent within one-to-five years. See Jim VandeHei and Mike Allen, "Behind the Curtain: A White-Collar Bloodbath," Axios, May 28, 2025, https://www.axios.com/2025/05/28/ai-jobs-white-collar-unemployment-anthropic. For Sam Altman's remarks that "entire classes of jobs will go away," see "OpenAI CEO Sam Altman Speaks at Federal Reserve Conference," C-SPAN, July 22, 2025, archived July 26, 2025, https://archive.org/details/CSPAN_20250726_231200_OpenAI_CEO_Sam_Altman_Speaks_at_Federal_Reserve_Conference.

3 "Artificial intelligence" is far older and broader than the technology this book discusses. It encompasses search algorithms and logic-based reasoning through machine learning, computer vision, and other techniques. For a review, see Stuart Russell and Peter Norvig, *Artificial Intelligence: A Modern Approach*, 4th ed. (Hoboken, NJ: Pearson, 2021). What changed in 2022 to 2023, and what drives the economic transformation we analyze here, is the rise of large language models (LLMs), which are neural networks trained on vast quantities of text that can generate, summarize, translate, and reason across domains with minimal task-specific instruction.

4 For evidence on the efficiency of using AI for academic research, see Keigo Kusumegi, Xinyu Yang, Paul Ginsparg, Mathijs de Vaan, Toby E. Stuart, and Yian Yin, "Scientific Production in the Era of Large Language Models," *Science* 390, no. 6779 (2025): 1240-43, https://doi.org/10.1126/science.adw3000.

5 For a vivid description of this historical episode, see Peter N. Stearns, *The Industrial Revolution in World History*, 4th ed. (Westview Press, 2013).

6 For a discussion of the economic consequences of the Industrial Revolution, see Robert C. Allen, *The British Industrial Revolution in Global Perspective* (Cambridge University Press, 2009).

7 For a thorough discussion of the evidence and the role of worker bargaining power and institutions in attaining broad prosperity, see Daron Acemoglu and Simon Johnson, *Power and Progress: Our Thousand-Year Struggle over Technology and Prosperity* (PublicAffairs, 2023).

8 See Leila Bengali, Robert G. Valletta, and Cindy Zhao, "Explaining Stagnation in the College Wage Premium," Federal Reserve Bank of San Francisco Working Paper 2025-01 (2025).

9 Lisa Camner McKay, "What Happened to the College Wage Premium?" Federal Reserve Bank of Minneapolis, September 8, 2025.

10 See David H. Autor, Lawrence F. Katz, and Melissa S. Kearney, "The Polarization of the US Labor Market," *American Economic Review* 96, no. 2 (2006): 189-194.

11 The cost of hiring a human includes the wage, which may go down as the technology improves, but also many fixed costs including supervising the worker, and regulatory and liability costs.

12 To be precise, as we discuss in our March 20, 2026 working paper "Weak Bundle, Strong Bundle: How AI Redraws Job Boundaries," the autonomy threshold may depend on worker skill. Workers who are very good at a particular task may continue doing it longer with machine help than those who are not great at it.

13 For instance, job postings for automation-prone occupations on online freelancing platforms, such as writing, software and app development, and engineering, declined relative to manual-intensive occupations after the introduction of ChatGPT. Ozge Demirci, Jonas Hannane, and Xinrong Zhu, "Who Is AI Replacing? The Impact of Generative AI on Online Freelancing Platforms," *Management Science* (2025), https://doi.org/10.1287/mnsc.2024.05420.

14 These two configurations, Robots Above and Robots Below, are studied in Enrique Ide and Eduard Talamàs, "Artificial Intelligence in the Knowledge Economy," *Journal of Political Economy* 133, no. 12 (2025): 3762-3800. Relying on the knowledge-economy framework is studied

in Luis Garicano, "Hierarchies and the Organization of Knowledge in Production," *Journal of Political Economy* 108, no. 5 (2000): 874-904; and Luis Garicano and Esteban Rossi-Hansberg, "Organization and Inequality in a Knowledge Economy," *Quarterly Journal of Economics* 121, no. 4 (2006): 1383-1435.

15 Boris Cherny, post on X, February 14, 2026, https://x.com/i/status/2022762422302576970.

16 William D. Nordhaus, "Do Real-Output and Real-Wage Measures Capture Reality? The History of Lighting Suggests Not," *Quarterly Journal of Economics* 112, no. 3 (1997): 937-975.

17 Daniel Gross, "US Farms Still Feed the World, but Farm Jobs Dwindle," Strategy+Business, August 2, 2016, https://www.strategy-business.com/blog/US-Farms-Still-Feed-the-World-But-Farm-Jobs-Dwindle.

18 Kathleen Kassel, "Agriculture and Its Related Industries Provide 10.4 Percent of US Employment," Economic Research Service, US Department of Agriculture, November 3, 2023, https://www.ers.usda.gov/data-products/chart-gallery/chart-detail?chartId=58282.

19 See William Stanley Jevons, *The Coal Question: An Inquiry Concerning the Progress of the Nation, and the Probable Exhaustion of Our Coal-Mines* (Macmillan, 1865).

20 In a thread on X, https://x.com/karpathy/status/1886192184808149383.

21 Writing on X in February 2025, https://x.com/karpathy/status/1886192184808149383?s=20.

22 Brian Caulfield, "NVIDIA CEO: Every Country Needs Sovereign AI," Nvidia Blog, February 12, 2024, https://blogs.nvidia.com/blog/world-governments-summit.

23 See for example Kiran Tomlinson, Sonia Jaffe, Will Wang, Scott Counts, and Siddharth Suri, "Working with AI: Measuring the Applicability of Generative AI to Occupations," arXiv preprint arXiv:2507.07935 (2025).

24 Erik Brynjolfsson, Danielle Li, and Lindsey R. Raymond, "Generative AI at Work," *Quarterly Journal of Economics* 140, no. 2 (May 2025): 889-942, https://doi.org/10.1093/qje/qjae044.

25 For more details, see Jie Gong, Jin Li, Lilian Xu, and Leila Zhang, "The C-Smart Agent: A New Service Model at Yum China," nano case, HKU Centre for AI, Management and Organisation, University of Hong Kong, February 23, 2026, https://camo.hku.hk/nanocase-thecsmartagent/.

26 Pranshu Verma, Gerrit De Vynck, "ChatGPT Took Their Jobs. Now They Walk Dogs and Fix Air Conditioners." *The Washington Post*, June 2, 2023, https://www.washingtonpost.com/technology/2023/06/02/ai-taking-jobs/.

27 In benchmarks that compare agent performance to human time-to-complete, success rates fall sharply as tasks stretch from minutes to hours. Researchers summarize this as a time horizon: the human task length an agent can complete with 50 percent success. On structured software task sets, the 50 percent time horizon for frontier agents is now measured in hours and has been rising quickly. Extrapolations do suggest much longer horizons within a few years, but the estimates vary sharply by domain and by how "clean" the tasks are. When METR regresses the success rate on the logarithm of human time-to-complete, the fit is very tight ($R^2 \approx 0.83$), implying that task duration alone explains most of the variation in performance across tasks. METR, "Measuring AI Ability to Complete Long Tasks," technical report / preprint (March 2025); METR, "HCAST: Human-Calibrated Autonomy Software Tasks," technical report (2025).

28 Fabrizio Dell'Acqua, Edward McFowland III, Ethan Mollick, Hila Lifshitz, Katherine C. Kellogg, Saran Rajendran, Lisa Krayer, François Candelon, Karim R. Lakhani, "Navigating the Jagged Technological Frontier: Field Experimental Evidence of the Effects of Artificial Intelligence on Knowledge Worker Productivity and Quality," *Organization Science* 37, no. 2 (2026): 403-423.

29 A more fundamental limitation has been claimed by Yann LeCun, one of the fathers of neural networks, who has argued that language models lack a world model. They can only manipulate words, and hence they predict where human words would lead, but they do not accumulate knowledge about the world the way humans do. Davos AI Summit, Imagination in Action, "Why LLMs Will Not Lead to AGI | Yann LeCun," YouTube video, 29:11, January 27, 2026, https://www.youtube.com/watch?v=5PQtJxd4U0M

30 Jason Wei, "Asymmetry of Verification and Verifier's Rule," author's blog, 2025, https://www.jasonwei.net/blog/asymmetry-of-verification-and-verifiers-law.

31 Blake Crosley, "World Models Race 2026: How LeCun, DeepMind, and World Labs Are Redefining the Path to AGI," Introl, January 3, 2026, https://introl.com/blog/world-models-race-agi-2026.

32 "Anthropic CEO Dario Amodei: AI's Potential, OpenAI Rivalry, GenAI Business, Doomerism," YouTube video, https://www.youtube.com/watch?v=mYDSSRS-B5U.

33 Mrinank Sharma, Meg Tong, Tomasz Korbak, et al., "Towards Understanding Sycophancy in Language Models," arXiv:2310.13548 (2023; updated 2025).

34 Jiseung Hong, Grace Byun, Seungone Kim, and Kai Shu, "Measuring Sycophancy of Language Models in Multi-Turn Dialogues," in Findings of the Association for Computational Linguistics: EMNLP 2025, Association for Computational Linguistics (2025): 2239–59, https://doi.org/10.18653/v1/2025.findings-emnlp.121.

35 Avneet Kaur, "Echoes of Agreement: Argument Driven Sycophancy in Large Language Models," in Findings of the Association for Computational Linguistics: EMNLP 2025, Association for Computational Linguistics (2025): 22803–12, https://doi.org/10.18653/v1/2025.findings-emnlp.1241.

36 Fortinet, "AI Security: Addressing Emerging Generative AI Risks," Fortinet CyberGlossary, accessed March 5, 2026, https://www.fortinet.com/resources/cyberglossary/ai-security.

37 Rafe Rosner-Uddin, "Amazon Service Was Taken Down by AI Coding Bot," *Financial Times*, February 20, 2026.

38 Lee Gray, "The History of Operatorless Elevators: Traffic Control Systems (Part One)," Elevator World, August 1, 2023, https://elevatorworld.com/article/the-history-of-operatorless-elevators-traffic-control-systems-part-one/.

39 "An elevator strike could effectively paralyze the economy." Building operating costs doubled "from twenty-six cents to fifty cents per square foot between 1952 and 1953." See Grace Palladino, "When Militancy Isn't Enough: The Impact of Automation on New York City Building Service Workers, 1934-1970," *Labor History* 28, no. 2 (1987): 196–220, https://doi.org/10.1080/00236568700890111. See also Steve Henn, "Remembering When Driverless Elevators Drew Skepticism," Planet Money, NPR, July 31, 2015, https://www.npr.org/2015/07/31/427990392/remembering-when-driverless-elevators-drew-skepticism.

40 Maya Yang, "China's Dancing Robots: How Worried Should We Be?," The Guardian, February 18, 2026, https://www.theguardian.com/world/2026/feb/18/china-dancing-humanoid-robots-festival-show.

41 On writing with ChatGPT 3.5, see Shakked Noy and Whitney Zhang, "Experimental Evidence on the Productivity Effects of Generative Artificial Intelligence," *Science* 381, no. 6654 (2023), https://doi.org/10.1126/science.adh2586. On customer service agents using GPT-3, see Erik Brynjolfsson, Danielle Li, and Lindsey R. Raymond, "Generative AI at Work," *Quarterly Journal of Economics* 140, no. 2 (May 2025): 889-942. On software, using 2022-2023 GitHub Copilot, see Kevin Zheyuan Cui, Mert Demirer, Sonia Jaffe, Leon Musolff, Sida Peng, and Tobias Salz, "The Effects of Generative AI on High-Skilled Work: Evidence from Three Field Experiments with Software Developers," *Management Science*, published online February 27, 2026.

42 For media coverage of Block's staff downsizing, see Ramishah Maruf, "Block Lays Off Nearly Half Its Staff Because of AI. Its CEO Said Most Companies Will Do the Same," CNN, February 27, 2026, https://edition.cnn.com/2026/02/26/business/block-layoffs-ai-jack-dorsey. For Dorsey's post on X, see https://x.com/jack/status/2027129697092731343?s=20.

43 See "The List of Major Companies Laying Off Staff This Year Includes Verizon, IBM, Amazon, Starbucks, American Airlines, and More," December 17, 2025, *Business Insider*, https://www.businessinsider.com/recent-company-layoffs-laying-off-workers-2025.

44 Leland D. Crane and Paul E. Soto, "AI and Coder Employment: Compiling the Evidence, Finance, and Economics," Discussion Series no. 2026-018, Board of Governors of the Federal Reserve System (2026).

45 Allen & Overy press release, February 2023; David Wakeling, quoted in Caroline Hill, "Allen & Overy Breaks the Internet (and New Ground) with Co-Pilot Harvey," Legal IT Insider, February 16, 2023. For Wakeling's quote, see Chris Stokel-Walker, "Generative AI Is Coming for the Lawyers." *WIRED*, February 21, 2023.

46 Harvey raised $160 million in a Series F round led by Andreessen Horowitz in December 2025, valuing the company at roughly $8 billion. Harvey, "Andreessen Horowitz Leads $160M Investment in Harvey," press release, December 4, 2025, https://www.harvey.ai/blog/andreessen-horowitz-leads-dollar160m-investment-in-harvey.

47 Jonathan H. Choi, Amy B. Monahan, and Daniel Schwarcz, "Lawyering in the Age of Artificial Intelligence," *Minnesota Law Review* 109 (2024): 147.

48 The company describes itself this way: "Hims & Hers is a digital health platform that connects people to licensed healthcare providers and, if prescribed, personalized treatment plans." It is listed on the New York Stock Exchange under ticker HIMS. See Hims & Hers, "About Hims & Hers," https://www.hims.com/about/the-company.

49 The description is based on social media posts in Chinese and the following report: "Baidu Robotaxis Draw Complaints from Human Drivers as Service Gains Popularity," *South China Morning Post*, July 10, 2024.

50 The video of his remarks has become viral: https://www.youtube.com/watch?v=2HMPRXstSvQ.

51 Ram Sivakumar, Brian Lue, and Shinjini Kundu, "FDA Approval of Artificial Intelligence and Machine Learning Devices in Radiology: A Systematic Review," *JAMA Network Open* 8, no. 11 (November 3, 2025), e2542338, https://doi.org/10.1001/jamanetworkopen.2025.42338.

52 Marty Stempniak, "Radiology Rises to the No. 3 Highest Paid Specialty, Surpassing Cardiology," *Radiology Business*, April 14, 2025, https://radiologybusiness.com/topics/healthcare-management/radiologist-salary/radiology-rises-no-3-highest-paid-specialty-surpassing-cardiology-medscape.

53 The three of us carefully and formally analyze the distinction between weak and strong bundles. See Luis Garicano, Jin Li, and Yanhui Wu, "Weak Bundle, Strong Bundle: How AI Redraws Job Boundaries," CEPR Press, no. DP21453, 2026.

54 Gary S. Becker and Kevin M. Murphy, "The Division of Labor, Coordination Costs, and Knowledge," *The Quarterly Journal of Economics* 107, no. 4 (1992): 1137-60.

55 Encyclopaedia Britannica, "Challenger Disaster," accessed March 2026, https://www.britannica.com/event/Challenger-disaster.

56 Michael Kremer, "The O-Ring Theory of Economic Development," *The Quarterly Journal of Economics* 108, no. 3 (1993): 551-75.

57 For the details of this research, see Fabrizio Dell'Acqua, Charles Ayoubi, Hila Lifshitz, Raffaella Sadun, Ethan Mollick, Lilach Mollick, Yi Han, et al. "The Cybernetic Teammate: A Field Experiment on Generative AI Reshaping Teamwork and Expertise," Harvard Business School Working Paper, no. 25-043, March 2025.

58 Nan Jia, Xueming Luo, Zheng Fang, and Chengcheng Liao, "When and How Artificial Intelligence Augments Employee Creativity," *Academy of Management Journal* 67 (2023), https://doi.org/10.5465/amj.2022.0426.

59 Armen A. Alchian and Harold Demsetz, "Production, Information Costs, and Economic Organization," *American Economic Review* 62, no. 5 (1972): 777-95.

60 David J. Deming, "The Growing Importance of Social Skills in the Labor Market," *Quarterly Journal of Economics* 132, no. 4 (2017): 1593-1640.

61 Catherine Weinberger, "The Increasing Complementarity between Cognitive and Social Skills," *Review of Economics and Statistics* 96, no. 5 (2014): 849-861.

62 Ben Weidmann and David Deming, "Team Players: How Social Skills Improve Team Performance," *Econometrica* 89, no. 6 (2021): 2637-57.

63 See US Bureau of Labor Statistics, Employment Projections, 2024-34, published August 2025.

64 "6-Month-Old, Solo-Owned Vibe Coder Base44 Sells to Wix for $80M Cash." TechCrunch, June 20, 2025, https://techcrunch.com/2025/06/18/6-month-old-solo-owned-vibe-coder-base44-sells-to-wix-for-80m-cash/.

65 Economic theory of management by exception is introduced in Luis Garicano, "Hierarchies and the Organization of Knowledge in Production," *Journal of Political Economy* 108, no. 5 (2000): 874-904, and, in an equilibrium context, by Luis Garicano and Esteban Rossi-Hansberg, "Organization and Inequality in a Knowledge Economy," *The Quarterly Journal of Economics* 121, no. 4 (2006): 1383-1435.

66 Nicholas G. Otis, Rowan P. Clarke, Solène Delecourt, David Holtz, and Rembrand Koning, "The Uneven Impact of Generative AI on Entrepreneurial Performance," OSF Preprints (2023), https://doi.org/10.31219/osf.io/hdjpk.

67 Thinkific, "How 'Miss Excel' Launched a Six-Figure Online Course Business on TikTok," July 14, 2025, https://www.thinkific.com/case-studies/miss-excel-online-course-tiktok-launch/.

68 For an interesting analysis about the shapes of the talent and income distributions, see Xavier Gabaix, "Power Laws in Economics and Finance," *Annual Review of Economics* 1 (2009): 255–94, https://doi.org/10.1146/annurev.economics.050708.142940.

69 The classic analysis of "Superstar" markets is by Sherwin Rosen, "The Economics of Superstars," *The American Economic Review* 71, no. 5 (1981): 845-858.

70 For an excellent survey on a large number of studies on this topic, see Anthony B. Atkinson, Thomas Piketty, and Emmanuel Saez, "Top Incomes in the Long Run of History," *Journal of Economic Literature* 49, no. 1 (2011): 3-71.

71 CNBC, "Sam Altman Says Meta Offered OpenAI Staff $100 Million Bonuses, As Mark Zuckerberg Ramps Up AI Poaching Efforts." June 18, 2025,_ https://www.cnbc.com/2025/06/18/sam-altman-says-meta-tried-to-poach-openai-staff-with-100-million-bonuses-mark-zuckerberg.html.

72 Lenny Rachitsky, "The Battle for AI Talent: Anthropic Cofounder On Why Mission Matters More Than Money," Lenny's Podcast, July 2025.

73 Luis Aguiar and Joel Waldfogel, "Platforms, Promotion, and Product Discovery: Evidence from Spotify Playlists," NBER Working Paper 24713 (2018), https://www.nber.org/papers/w24713.

74 Daron Acemoglu and Pascual Restrepo, "Automation and New Tasks: How Technology Displaces and Reinstates Labor," *Journal of Economic Perspectives* 33, no. 2 (2019): 3-30.

75 A form of this argument was first made by David Autor, "Applying AI to Rebuild Middle Class Jobs," NBER Working Paper 32140 (2024), https://doi.org/10.3386/w32140.

76 For more details about this research, see Kyogo Kanazawa et al., "AI, Skill, and Productivity: The Case of Taxi Drivers," NBER Working Paper 30612 (October 2022), https://doi.org/10.3386/w30612.

77 For more details about the adoption and use of AI at KFC China, see Jie Gong, Jin Li, and Aiken Yuen, "Scaling Managers with AI at KFC China," HKU Centre for AI, Management and Organisation (CAMO), January 15, 2026, https://camo.hku.hk/nano-case-3kfcchina/.

78 CVS Health, "MinuteClinic® Primary Care Expands Access Amid Shortage of Primary Care Physicians," CVS Health, September 17, 2025, https://www.cvshealth.com/news/company-news/minuteclinic-primary-care-expands-access-amid-shortage.html.

79 US Bureau of Labor Statistics, "Nurse Anesthetists, Nurse Midwives, and Nurse Practitioners," *Occupational Outlook Handbook*, accessed March 4, 2026, https://www.bls.gov/ooh/healthcare/nurse-anesthetists-nurse-midwives-and-nurse-practitioners.htm.

80 For discussions along this line, see Brian Consiglio, "States That Empower Advanced Practice Registered Nurses Have Better Health Outcomes, New Study Finds," Sinclair School of Nursing, University of Missouri, July 7, 2025, https://nursing.missouri.edu/states-that-empower-advanced-practice-registered-nurses-have-better-health-outcomes-new-study-finds/.

81 See International Energy Agency, *Electricity 2024: Analysis and Forecast to 2026* (IEA, 2024), 31-33, https://www.iea.org/reports/electricity-2024.

82 Associated Builders and Contractors, "Construction Workforce Shortage Tops Half a Million in 2024, Says ABC," January 31, 2024, https://www.abc.org/News-Media/News-Releases/construction-workforce-shortage-tops-half-a-million-in-2024-says-abc.

83 Nina Dale, "Data Centers Drive Skilled Trades Hiring Boom," CRE Daily, December 2, 2025, https://credaily.com/briefs/data-centers-drive-skilled-trades-hiring-boom/.

84 David Autor and Neil Thompson, "Expertise," *Journal of the European Economic Association* 23, no. 4 (August 2025): 1203-71, https://doi.org/10.1093/jeea/jvaf023.

85 A detailed description of the Colossus project can be found at https://x.ai/colossus.

86 For a beautiful analysis of the London housing problem, see John Burn-Murdoch, "How London Unwittingly Killed Housebuilding," *Financial Times*, February 13, 2026, https://www.ft.com/content/05faba7d-cb8e-470b-b35f-271517b99d92.

87 R. F. Meyer, M. G. Rukstad, P. J. Coughlan, and S. A. Jansen, "DaimlerChrysler Post-Merger Integration (A)," Harvard Business School Case 703-417, September 2002, revised December 2005.

88 See F. A. Hayek, "The Use of Knowledge in Society," *The American Economic Review* 35, no. 4 (1945): 519-30, http://www.jstor.org/stable/1809376.

89 Kenneth J. Arrow, *The Limits of Organization* (W. W. Norton & Company, 1974).

90 See Oliver E. Williamson, "Credible Commitments: Using Hostages to Support Exchange," *American Economic Review* 73, no. 4 (1983): 519–540.

91 See Oliver Hart and John Moore, "Property Rights and the Nature of the Firm," *Journal of Political Economy* 98, no. 6 (1990): 1119-1158.

92 See for example Naomi Fulop et al., "Process and Impact of Mergers of NHS Trusts: Multicentre Case Study and Management Cost Analysis," *BMJ* 325, no. 7358 (August 3, 2002): 246, https://doi.org/10.1136/bmj.325.7358.246.

93 To encourage people to communicate truthfully, this must be incentive compatible. A long literature explores this problem, starting from Leo Hurwicz's seminal contribution "On Informationally Decentralized Systems," in *Decision and Organization: A Volume in Honor of Jacob Marschak*, ed. C. B. McGuire and Roy Radner (North-Holland, 1972).

94 Roger B. Myerson and Mark A. Satterthwaite, "Efficient Mechanisms for Bilateral Trading," *Journal of Economic Theory* 29, no. 2 (1983): 265-81. The theorem applies strictly to bilateral trade, but the logic generalizes. Whenever parties hold private information about their own costs, values, or intentions, no mechanism can guarantee honest reporting, efficient allocation, and voluntary participation simultaneously. Inside organizations, where teams compete for budgets, talent, and priority, this problem is ubiquitous.

95 For news coverage of the event, see "Sonos CEO Patrick Spence Steps Down after App Update Debacle," CNBC, January 13, 2025, https://www.cnbc.com/2025/01/13/sonos-ceo-patrick-spence-steps-down-after-app-update-debacle.html. For a more in-depth report, see Mark Gurman, "Sonos CEO Leaving after Botched App Revamp Led to Customer Revolt," Bloomberg, January 13, 2025, reprinted in *Fortune*.

96 See Gillian K. Hadfield and Andrew Koh, "An Economy of AI Agents," arXiv:2509.01063 (2025).

97 See Allan E. Lind and Tom R. Tyler, *The Social Psychology of Procedural Justice* (Plenum Press, 1988).

98 "BRAC Commission Is Established to Close US Military Bases," EBSCO Research Starters, https://www.ebsco.com/research-starters/history/brac-commission-established-close-us-military-bases.

99 Many practices in organizations appear inefficient but serve the purpose of establishing legitimacy. The seminal contribution on this topic is John W. Meyer and Brian Rowan, "Institutionalized Organizations: Formal Structure as Myth and Ceremony," *American Journal of Sociology* 83, no. 2 (1977): 340-63.

100 Henry Chandonnet, "Meta AI Alignment Director Shares Her OpenClaw Email-Deletion Nightmare," Business Insider, February 24, 2026, https://www.businessinsider.com/meta-ai-alignment-director-openclaw-email-deletion-2026-2.

101 See Letian Zhang, "The Changing Role of Managers," *American Journal of Sociology* 129, no. 2 (September 2023): 439-484.

102 See McKinsey & Company, "Stop Wasting Your Most Precious Resource: Middle Managers," March 10, 2023, https://www.mckinsey.com/capabilities/people-and-organizational-performance/our-insights/stop-wasting-your-most-precious-resource-middle-managers.

103 Luis Garicano and Paul Heaton, "Information Technology, Organization, and Productivity in the Public Sector: Evidence from Police Departments," *Journal of Labor Economics* 28, no. 1 (January 2010): 167-201, https://doi.org/10.1086/649844.

104 See table 4 and related discussion (on reporting effects and paperwork) and table 7 and discussion (on Compstat as a bundle) in Luis Garicano and Paul Heaton, "Information Technology, Organization, and Productivity in the Public Sector."

105 Microsoft, "Microsoft and LinkedIn release the 2024 Work Trend Index on the State of AI at Work," Microsoft Source, May 8, 2024, accessed March 1, 2026, https://news.microsoft.com/source/2024/05/08/microsoft-and-linkedin-release-the-2024-work-trend-index-on-the-state-of-ai-at-work/.

106 Rebecca M. Henderson and Kim B. Clark, "Architectural Innovation: The Reconfiguration of Existing Product Technologies and the Failure of Established Firms," *Administrative Science Quarterly* 35, no. 1 (March 1990): 9-30, https://doi.org/10.2307/2393549.

107 Paul Milgrom and John Roberts, "The Economics of Modern Manufacturing: Technology, Strategy, and Organization," *American Economic Review* 80, no. 3 (June 1990): 511-28.

108 Casey Ichniowski, Kathryn Shaw, and Giovanna Prennushi, "The Effects of Human Resource Management Practices on Productivity: A Study of Steel Finishing Lines," *American Economic Review* 87, no. 3 (1997): 291-313.

109 James G. March, "Exploration and Exploitation in Organizational Learning," *Organization Science* 2, no. 1 (February 1991): 71-87, https://doi.org/10.1287/orsc.2.1.71.

110 Colleen Pope, "'Meaningful Use' of Electronic Health Records," *Health Policy Brief, Health Affairs*, August 24, 2010, https://www.healthaffairs.org/do/10.1377/hpb20100824.587990/; Office of the National Coordinator for Health Information Technology, "Adoption of Electronic Health Record Systems among US Non-Federal Acute Care Hospitals: 2008-2015," ONC Data Brief no. 35 (May 2016), https://healthit.gov/data/data-briefs/adoption-electronic-health-record-systems-among-us-non-federal-acute-care-1/.

111 Christine Sinsky et al., "Allocation of Physician Time in Ambulatory Practice: A Time and Motion Study in 4 Specialties," *Annals of Internal Medicine* 165, no. 11 (December 6, 2016): 753-760, https://doi.org/10.7326/M16-0961.

112 Paul A. David, "The Dynamo and the Computer: An Historical Perspective on the Modern Productivity Paradox," *American Economic Review* 80, no. 2 (May 1990): 355-361.

113 James Dow and Enrico Perotti, "Resistance to Change," Fondazione Eni Enrico Mattei Working Papers, Paper 437 (May 10, 2010), https://services.bepress.com/feem/paper437/.

114 Peter Moskos, interview by Santi Ruiz, "How to Beat Crime in New York City," Statecraft (podcast transcript), April 3, 2025, https://www.statecraft.pub/p/how-to-fix-crime-in-new-york-city.

115 J. D. Tuccille, "Jack Maple: Betting on Intelligence," *Government Technology*, August 12, 2010, https://www.govtech.com/magazines/gt/jack-maple-betting-on-intelligence.html.

116 Peter Moskos, "Reducing Crime in NYC," What Happens Next in 6 Minutes (podcast transcript), August 16, 2025, https://www.whathappensnextin6minutes.com/p/reducing-crime-in-nyc.

117 William J. Bratton, *Zero Tolerance: Policing a Free Society* (Civitas, 1998), p. 59, https://civitas.org.uk/content/files/cw351.pdf.

118 John Timoney, remarks in "Police Leaders at PERF/BJA Meeting Discuss CompStat," Subject to Debate (Police Executive Research Forum), March/April 2011, 3, https://perf.memberclicks.net/assets/docs/Subject_to_Debate/Debate2011/debate_2011_marapr.pdf.

119 Elena Alfaro, Antonio Cabrales, José Elías Durán Roa, Luis Garicano, Isabel Pérez del Caño, Toni Roldán Monés, and Guillermo Vieira de Santiago, "The Hidden Demand for AI Inside Your Company," *Harvard Business Review*, April 14, 2026, https://hbr.org/2026/04/the-hidden-demand-for-ai-inside-your-company.

120 Boston Consulting Group, "AI Adoption in 2024: 74% of Companies Struggle to Achieve and Scale Value," press release, October 24, 2024, accessed March 1, 2026, https://www.bcg.com/press/24october2024-ai-adoption-in-2024-74-of-companies-struggle-to-achieve-and-scale-value.

121 Stanford's AI Index estimates corporate AI investment at $252.3 billion in 2024. See Stanford Institute for Human-Centered Artificial Intelligence, *Artificial Intelligence Index Report* 2025 (Stanford University, 2025), https://hai-production.s3.amazonaws.com/files/hai_ai_index_report_2025.pdf.

122 Joseph A. Schumpeter, *Capitalism, Socialism, and Democracy*, 5th ed. (George Allen & Unwin, 1976; reprinted by Taylor & Francis e-Library, 2003).

123 George Johnson, "To Test a Powerful Computer, Play an Ancient Game," *The New York Times*, July 29, 1997.

124 Samuel Gibbs, "AlphaZero AI Beats Champion Chess Program after Teaching Itself in Four Hours," *The Guardian*, December 7, 2017.

125 Colin McGourty, "Carlsen Stuns Caruana to Win 21st World Championship Title," Chess.com, 2026.

126 Allison Pohle, "AI Can't Touch These Skilled Trade Jobs. If Only Enough Humans Would Fill Them," *The Wall Street Journal*, January 14, 2026.

127 US Bureau of Labor Statistics, "Wholesale and Manufacturing Sales Representatives," *Occupational Outlook Handbook*, accessed March 5, 2026, https://www.bls.gov/ooh/sales/wholesale-and-manufacturing-sales-representatives.htm.

128 Alex Imas and Kristóf Madarász, "Superiority-Seeking and the Preference for Exclusion," *Review of Economic Studies* 91, no. 4 (July 2024), https://doi.org/10.1093/restud/rdad079.

129 Graelin Mandel and Alex Imas, "Art and the Machine: Why People Devalue AI-Generated Creative Work," SSRN, February 25, 2026, https://doi.org/10.2139/ssrn.6302659.

130 Katherine Laidlaw, "How to Win the Hermès Game," The Hustle, May 10, 2024, https://thehustle.co/originals/how-to-win-the-herm%C3%A8s-game.

131 Xueming Luo, Siliang Tong, Zheng Fang, and Zhe Qu, "Frontiers: Machines vs. Humans: The Impact of AI Chatbot Disclosure on Customer Purchases," *Marketing Science* 38, no. 6 (2019): 937-47, https://doi.org/10.1287/mksc.2019.1192.

132 Similarly, Nicole Castelo, Johannes Boegershausen, Christian Hildebrand, and Andrea P. Henkel demonstrate that consumers rate the same service output more negatively when they believe it was produced by a bot rather than a human. They attribute this to a perception that companies deploy bots solely to cut costs at the expense of customer experience. They note this effect can be mitigated if the company shares efficiency gains with customers, for example through price discounts. Castelo et al., "Understanding and Improving Consumer Reactions to Service Bots," *Journal of Consumer Research* 50, no. 4 (2023): 848-63, https://doi.org/10.1093/jcr/ucad023.

133 Dirk Petzold, "Human-Made Design Authenticity Answers the Growing Demand for Genuine Human Connection," We and the Color (blog), https://weandthecolor.com/human-made-design-authenticity-answers-the-growing-demand-for-genuine-human-connection/207475.

134 Johan Brenner, "Vivino: From a Free Wine Scanning App to the World's Largest Wine Marketplace," Creandum (blog), February 4, 2021, https://blog.creandum.com/vivino-from-a-free-wine-scanning-app-to-the-worlds-largest-wine-marketplace-e7c65a3bb65c.

135 Custom Market Insights, "Global Live Music Market Size, Trends, Share 2025–2034," https://www.custommarketinsights.com/report/live-music-market/.

136 Live Nation Entertainment, "Live Nation Entertainment Reports Full Year and Fourth Quarter 2024 Results," February 20, 2025, https://newsroom.livenation.com/news/live-nation-entertainment-reports-full-year-and-fourth-quarter-2024-results/.

137 Paul Gatza, "US Craft Brewery Count Reaches 2000," Brewers Association, June 2012, https://www.brewersassociation.org/brewing-industry-updates/u-s-craft-brewery-count-reaches-2000/.

138 Brewers Association, "Brewers Association Reports 2024 US Craft Brewing Industry Figures," May 2025, https://www.brewersassociation.org/association-news/brewers-association-reports-2024-u-s-craft-brewing-industry-figures/. See also Sara Gherghelas, "NFT Art's Shocking Collapse: From $2.9 Billion Boom to $23.8 Million Bust—What Went Wrong?," DappRadar (blog), 2025, https://dappradar.com/blog/nft-arts-shocking-collapse-from-2-9-billion-boom-to-23-8-million-bust-what-went-wrong.

139 Nikhil Batra, "National Handloom Day 2024: PM Modi Showed AI Advancement in Handloom and Textiles," Jagran Josh, August 7, 2024, https://www.jagranjosh.com/general-knowledge/national-handloom-day-1723010364-1.

140 "Wabi-Sabi in Modern Homes: Embracing Imperfection Through Japanese Design," *The Rikumo Journal*, December 1, 2025, https://journal.rikumo.com/journal/2025/12/1/wabi-sabi-in-modern-homes-embracing-imperfection-through-japanese-design.

141 Alex Imas, "What Will Be Scarce?," Ghosts of Electricity, April 2026. Imas argues that as automation lowers the cost of commodity production, scarcity shifts toward relational goods and services whose value depends on human presence, provenance, care, status, and social meaning.

142 Abram Brown, "Beeple NFT Sells for $69.3 Million, Becoming Most-Expensive Ever," *Forbes*, March 11, 2021, https://www.forbes.com/sites/abrambrown/2021/03/11/beeple-art-sells-for-693-million-becoming-most-expensive-nft-ever/.

143 See Walter Benjamin, "The Work of Art in the Age of Mechanical Reproduction." In *Illuminations*, edited by Hannah Arendt, translated by Harry Zohn (Schocken Books, 1968), pp. 217-52.

144 Sara Gherghelas, "NFT Art's Shocking Collapse: From $2.9 Billion Boom to $23.8 Million Bust—What Went Wrong?" DappRadar (blog), 2025, https://dappradar.com/blog/nft-arts-shocking-collapse-from-2-9-billion-boom-to-23-8-million-bust-what-went-wrong.

145 Matt Medved, "Digital Art Isn't Dead—But It Is at a Crossroads," NFT Now, April 10, 2025, https://nftnow.com/newsletter/now-newsletter-digital-art-isnt-dead-but-it-is-at-a-crossroads/.

146 William J. Baumol and William G. Bowen, *Performing Arts: The Economic Dilemma* (Twentieth Century Fund, 1966).

147 Diego Comin, Danial Lashkari, and Martí Mestieri, "Structural Change with Long-Run Income and Price Effects," *Econometrica* 89, no. 1 (2021): 311-74. I owe this reference and the argument in this paragraph to Alex Imas. See his excellent article, "What Will Be Scarce? The Economics of Structural Change and the Post-Commodity Future of Work," Substack, https://aleximas.substack.com/p/what-will-be-scarce.

148 For a serious discussion of Baumol's cost disease, see William D. Nordhaus, "Baumol's Diseases: A Macroeconomic Perspective," *B.E. Journal of Macroeconomics* 8, no. 1 (2008): 1-39.

149 Regarding Roy Lee's suspension and Cluely's seed funding, see Charles Rollet, "Columbia Student Suspended over Interview Cheating Tool Raises $5.3M to 'Cheat on Everything,'" TechCrunch, April 21, 2025.

150 The demonstration of the AI tool being used on a dinner date is featured in the promotional video: Cluely, "Cluely (Cheating Tool for Literally Everything)," YouTube, April 20, 2025, https://www.youtube.com/watch?v=Rz3LD7u2KX8.

151 On the "rat race" and adverse selection in law-firm hours, see Renée M. Landers, James B. Rebitzer, and Lowell J. Taylor, "Rat Race Redux: Adverse Selection in the Determination of Work Hours in Law Firms," *American Economic Review* 86, no. 3 (1996): 329-48.

152 For research on dating-market game theory and signaling, see Paul Oyer, *Everything I Ever Needed to Know about Economics I Learned from Online Dating* (Harvard Business Review Press, 2014).

153 For the account of the eleven-minute standing ovation and the subsequent arrest of the factory director, see Aleksandr Solzhenitsyn, *The Gulag Archipelago, 1918–1956: An Experiment in Literary Investigation*, vol. 1, trans. Thomas P. Whitney (Harper & Row, 1973), part 1, chapter 1.

154 For the account of Sima Rangju's enforcement of military discipline to establish authority, see Sima Qian, "Sima Rangju Liezhuan" in *Shi Ji* (Records of the Grand Historian).

155 For Michael Spence's Nobel-winning signaling theory, see Michael Spence, *Market Signaling: Informational Transfer in Hiring and Related Screening Processes* (Harvard University Press, 1974).

156 For the research on the impact of AI writing assistance on hiring outcomes and resume evaluation, see Emma Wiles, Zanele T. Munyikwa, and John J. Horton, "Algorithmic Writing Assistance on Jobseekers' Resumes Increases Hires," NBER Working Paper 30886 (2023); and Bo Cowgill, Pablo Hernandez-Lagos, and Nataliya Wright, "Does AI Cheapen Talk? Theory and Evidence From Global Entrepreneurship and Hiring," Columbia Business School Research Paper, no. 4702114 (2024).

157 Regarding the proliferation of AI-generated mushroom foraging guides on Amazon and the subsequent safety warnings issued by mycological experts, see Dan Milmo, "Mushroom Pickers Urged to Avoid Foraging Books on Amazon That Appear to Be Written by AI," *The Guardian*, September 1, 2023.

158 For the fundamental research on the free-rider problem, see Mancur Olson, *The Logic of Collective Action: Public Goods and the Theory of Groups* (Harvard University Press, 1971).

159 For the research on how unobservable effort leads to moral hazard and the role of performance-based incentives in organizations, see Bengt Holmström, "Moral Hazard and Observability," *The Bell Journal of Economics* 10, no. 1 (1979): 74-91.

160 The 2001 Nobel Prize in Economic Sciences was awarded jointly to George A. Akerlof, A. Michael Spence, and Joseph E. Stiglitz for their pioneering analyses of markets with asymmetric information.

161 For the riverboat hauler example, see Steven N. S. Cheung, "The Contractual Nature of the Firm," *Journal of Law and Economics* 26, no. 1 (1983).

162 For the seminal formal analysis of how incentive contracts fail when an agent must allocate effort between measurable and hard-to-measure tasks, see Bengt Holmström and Paul Milgrom, "Multitask Principal-Agent

Analyses: Incentive Contracts, Asset Ownership, and Job Design," *Journal of Law, Economics, & Organization* 7, (1991): 24-52.

163 For a detailed analysis of how incentive structures led to the creation of unauthorized accounts at Wells Fargo, see Brian Tayan, "The Wells Fargo Cross-Selling Scandal," Stanford Closer Look Series (Stanford Graduate School of Business, January 2019).

164 For the study on how healthcare "report cards" led surgeons to decline difficult cases to protect their mortality statistics, see David Dranove, Daniel Kessler, Mark McClellan, and Mark Satterthwaite, "Is More Information Better? The Effects of 'Report Cards' on Health Care Providers," *Journal of Political Economy* 111, no. 3 (2003): 555-88.

165 For the analysis of the andon cord system and how Toyota's management principles empower workers to prioritize quality over immediate output, see Jeffrey K. Liker, *The Toyota Way: 14 Management Principles from the World's Greatest Manufacturer* (McGraw-Hill, 2004).

166 For the seminal research on relational contracts, see Clive Bull, "The Existence of Self-Enforcing Implicit Contracts," *Quarterly Journal of Economics* 102, no. 1 (1987): 147-59. For discussion of how relational contracts interact with formal contracts, see George Baker, Robert Gibbons, and Kevin J. Murphy, "Subjective Performance Measures in Optimal Incentive Contracts," *Quarterly Journal of Economics* 109, no. 4 (1994): 1125-56.

167 Regarding a government-commissioned report produced by Deloitte Australia that included fabricated references, see Justin Brake, "Major N.L. Healthcare Report Contains Errors Likely Generated by AI," *The Independent*, November 22, 2025.

168 For the research on how automated evaluation tools in software development can increase human moral hazard and reduce scrutiny of complex architectural risks, see Jong Sig Chung, "How Does Algorithmic Assistance Impact Human Performance in Collaboration? Evidence from Algorithm-Assisted Evaluation in Software Development," working paper.

169 For the report on authors embedding invisible AI instructions into arXiv papers to ensure positive feedback from AI-assisted reviewers, see Josh Taylor, "Scientists Reportedly Hiding AI Text Prompts in Academic Papers to Receive Positive Peer Reviews," *The Guardian*, July 14, 2025.

170 For the foundational research on how acquaintances and distant social connections are more likely to provide unique information and job opportunities than close friends, see Mark S. Granovetter, "The Strength of Weak Ties," *American Journal of Sociology* 78, no. 6 (1973): 1360-80.

171 Regarding the controversy over Jason Allen's award-winning AI-generated artwork at the Colorado State Fair and the process of prompt engineering and manual refinement involved, see Kevin Roose, "An AI-Generated Picture Won an Art Prize. Artists Aren't Happy," *The New York Times*, September 2, 2022; and Sarah Kuta, "Art Made With Artificial Intelligence Wins at State Fair," *Smithsonian Magazine*, September 6, 2022.

172 On Stanley Kubrick's directorial perfectionism during the production of *Eyes Wide Shut*, including the repetitive takes for simple actions and the record-breaking forty-six-week continuous shoot, see Nina Sterle, "They Shot This Movie for 400 Days Straight According to Guinness World Records," *Thought Catalog*, March 13, 2025.

173 For the definition of grit as a combination of passion and perseverance, see Angela Duckworth, *Grit: The Power of Passion and Perseverance* (Ebury Publishing, 2016).

174 For the story about the over-investment of labor into background details at Pixar, see Ed Catmull and Amy Wallace, *Creativity, Inc.: Overcoming the Unseen Forces That Stand in the Way of True Inspiration* (Random House, 2014).

175 For the account of Gustave Flaubert's meticulous writing process and his struggle to complete Madame Bovary, see Brian Patrick Eha, "Herald of Modernism," *City Journal*, Winter 2025.

176 For the story about the use of "popsicle sticks" to visualize labor trade-offs and resource management at Pixar, see Ed Catmull and Amy Wallace, *Creativity, Inc.: Overcoming the Unseen Forces That Stand in the Way of True Inspiration* (Random House, 2014).

177 For the account of Graham Greene's categorization of his books into "novels" and "entertainments" to manage artistic pressure and effort, see James Mustich, "On Reading Graham Greene: Page-Turning As Pilgrimage," *Medium*, April 24, 2020.

178 For the details of this experimental study, see Yvonne J. Chen, Jie Gong, Jin Li, and Zibo Zhao, "Better Technology, Worse Motivation: GenAI's Mediocrity Trap," HKU working paper.

179 For the research on the impact of early-2025 AI on the productivity of experienced open-source developers, which found that AI assistance increased completion time despite developer perceptions of speed, see Joel Becker, Nate Rush, Elizabeth Barnes, and David Rein, "Measuring the Impact of Early-2025 AI on Experienced Open-Source Developer Productivity," *METR* (July 2025).

180 For the research on how robotic surgery systems impede the training of medical residents and the story about the challenges faced by a surgeon named Kristen during her first independent procedure, see Matt Beane, *The Skill Code: How to Save Human Ability in an Age of Intelligent Machines* (Harper Business, 2024).

181 For reports on a tech executive shifting hiring standards because of AI and a founder replacing a large data science team with a single employee, see Kevin Roose, "For Some Recent Graduates, the AI Job Apocalypse May Already Be Here," *The New York Times*, 2025.

182 For reports on the incident involving marketing agencies reducing their demand for entry-level employees due to the influence of AI, see Lindsay Ellis and Katherine Bindley, "AI Is Wrecking an Already Fragile Job Market for College Graduates," *The Wall Street Journal*, July 29, 2025.

183 For reports on McKinsey & Company's use of AI to automate traditional analyst tasks, see Omar El Chmouri, "McKinsey Leans on AI to Make PowerPoints, Draft Proposals," Bloomberg, 2025.

184 For reports on the decline in new graduate hiring, see Asher Bantock, "The SignalFire State of Tech Talent Report 2025," SignalFire, 2025.

185 For data on unemployment-rate disparities between recent college graduates and other demographic groups, see Federal Reserve Bank of New York, "Unemployment Rates for Recent College Graduates versus Other Groups," https://www.newyorkfed.org/research/college-labor-market#--:explore:unemployment.

186 For the research on "seniority-biased technological change" and its impact on junior hiring versus expert productivity, see Seyed Mahdi Hosseini Maasoum and Guy Lichtinger, "Generative AI as Seniority-Biased Technological Change: Evidence from US Résumé and Job Posting Data," SSRN, August 31, 2025.

187 See Kenneth J. Arrow, "The Economic Implications of Learning by Doing," *The Review of Economic Studies* 29, no. 3 (1962): 155-73.

188 For the research on the "experience curve" and its application to unit cost reduction through cumulative production, see Bruce Henderson, "The Experience Curve," *BCG Perspectives*, 1968.

189 For the account of the "10,000-hour rule" as a metric for achieving expertise through deliberate practice, see Malcolm Gladwell, *Outliers: The Story of Success* (Little, Brown and Company, 2008).

190 For research on how early-career playing time in relegated soccer divisions leads to superior long-term earnings and career placement, see Jie Gong, Ang Sun, and Zhichao Wei, "Choosing the Pond: On-the-Job Experience and Long-Run Career Outcomes," *Management Science* 64, no. 2 (2018): 459-953.

191 For the research on the skill depreciation rate of Greek teachers awaiting formal assignments, see Michael Dinerstein, Rigissa Megalokonomou, and Constantine Yannelis, "Human Capital Depreciation and Returns to Experience," *American Economic Review* 112, no. 11 (2022): 3725-62.

192 For the research on the intergenerational transmission of musical style and the impact of mentor quality on the success of Western music composers since 1450, see Karol Jan Borowiecki, "Good Reverberations? Teacher Influence in Music Composition since 1450," *Journal of Political Economy* 130, no. 4 (2022): 991-1043.

193 For the research on the economic challenges of firm-sponsored training and the problem of employee poaching in competitive markets, see Gary S. Becker, *Human Capital: A Theoretical and Empirical Analysis with Special Reference to Education* (NBER, 1964).

194 For a general discussion on the "AI-Becker problem," see Luis Garicano, "The AI-Becker Problem: Who Will Train the Next Generation?," *Silicon Continent*, July 23, 2025.

195 Luis Garicano and Luis Rayo, "Training in the Age of AI: A Theory of Apprenticeship Viability," CEPR working paper, 2025.

196 For reports on AWS CEO Matt Garman's criticism of replacing junior employees with AI and the resulting long-term skill gap, see Lee Chong Ming, "Amazon Cloud Chief Says Replacing Junior Employees with AI Is 'One of the Dumbest Things I've Ever Heard,'" *Business Insider*, 2025.

197 Sam Altman has spoken of a future in which the cost of intelligence falls to nearly zero and the economy doubles every few years. See Sam Altman, "The Gentle Singularity," June 10, 2025, https://blog.samaltman.com/the-gentle-singularity. Dario Amodei has described a compressed century of scientific progress arriving in a decade. In this vision, the economy becomes a machine that optimizes itself. See Dario Amodei, "Machines of Loving Grace," October 2024, https://darioamodei.com/essay/machines-of-loving-grace.

198 Nobel Prize-winner Daron Acemoglu is a strong proponent of this view. The zero correlation between technological progress and productivity is called the *productivity paradox* in economics. This view is supported by the evidence that GDP per hour worked in the United States has grown at roughly 1.4 percent per year since 2005, barely half the rate of the post-war golden age.

199 Stripe cofounder John Collison posted on X on April 30, "The obvious near-term effect is reduced transaction costs within companies: shared context, systems of record, aligned incentives etc. But inter-company transaction costs also reduce sharply: agents are great at discovery, make it trivially easy to integrate; make contracting much more straightforward; agent-to-agent commerce. On net, we think second effect bigger in medium term: fewer people per firm, more output per firm, just more firms, and more coordination happening through market-like mechanisms." https://x.com/collision/status/2049974499278528952?

200 The view that firms exist as depositories of trust and as nexuses of long-term relationships has a long economic lineage, from Williamson's analysis of credible commitment to Baker, Gibbons, and Murphy's work on relational contracts. See Oliver E. Williamson, *The Mechanisms of Governance* (Oxford University Press, 1996); George Baker, Robert Gibbons, and Kevin J. Murphy, "Relational Contracts and the Theory of the Firm," *Quarterly Journal of Economics* 117, no. 1 (2002): 39-84; and

Robert Gibbons and John Roberts, eds. *The Handbook of Organizational Economics* (Princeton University Press, 2013), particularly the Gibbons-Henderson chapter on relational contracts.

201 The principle that innovations, or ideas, are the ultimate source of long-run growth is well stated in mainstream economics, starting Joseph A. Schumpeter, *Capitalism, Socialism, and Democracy* (Harper & Brothers, 1942) and the seminal work of Paul M. Romer inter alia in "Increasing Returns and Long-Run Growth," *Journal of Political Economy* 94, no. 5 (1986): 1002-1037. For a textbook treatment, see Daron Acemoglu, *Introduction to Modern Economic Growth* (Princeton University Press, 2009); Philippe Aghion and Peter Howitt, *The Economics of Growth* (MIT Press, 2009); and Charles I. Jones, *Introduction to Economic Growth*, 3rd ed. (W. W. Norton, 2013).

202 Charles I. Jones and Christopher Tonetti, "Past Automation and Future A.I.: How Weak Links Tame the Growth Explosion," Stanford working paper, March 29, 2026, https://web.stanford.edu/~chadj/JonesTonetti_Automation.pdf.

Index

C

X

Y

Z

www.ingramcontent.com/pod-product-compliance
Lightning Source LLC
LaVergne TN
LVHW100301210926
848480LV00003B/988